T0396719

Lecture Notes in Bioengineering

For further volumes:
http://www.springer.com/series/11564

Khin Wee Lai · Yan Chai Hum
Maheza Irna Mohamad Salim
Sang-Bing Ong · Nugraha Priya Utama
Yin Mon Myint · Norliza Mohd Noor
Eko Supriyanto

Advances in Medical Diagnostic Technology

 Springer

Khin Wee Lai
Biomedical Engineering Department
University Malaya
Kuala Lumpur
Malaysia

Yan Chai Hum
MIMOS Berhad
Technology Park Malaysia
Kuala Lumpur
Malaysia

Maheza Irna Mohamad Salim
Sang-Bing Ong, Ph.D. CBiol EurProBiol
Nugraha Priya Utama
Yin Mon Myint
Faculty of Biosciences and Medical
   Engineering
Universiti Teknologi Malaysia
Skudai-Johor
Malaysia

Norliza Mohd Noor
Department of Engineering
UTM Razak School of Engineering
   and Advanced Technology
Universiti Teknologi Malaysia
Kuala Lumpur
Malaysia

Eko Supriyanto
Faculty of Biosciences and Medical
   Engineering
IJN-UTM Cardiovascular Engineering
   Centre
Universiti Teknologi Malaysia
Skudai-Johor
Malaysia

ISSN 2195-271X   ISSN 2195-2728 (electronic)
ISBN 978-981-4585-71-2   ISBN 978-981-4585-72-9 (eBook)
DOI 10.1007/978-981-4585-72-9
Springer Singapore Heidelberg New York Dordrecht London

Library of Congress Control Number: 2014931386

Printed on acid-free paper

Springer is part of Springer Science+Business Media (www.springer.com)

Contents

Chapter 1
Ultrasonic Elastography and Breast Imaging

Yin Mon Myint, Khin Wee Lai, Maheza Irna Mohamad Salim, Yan Chai Hum and Nugraha Priya Utama

Abstract The elastography is based on the principles: (1) Tissue compression produces strain (displacement) within the tissue, and (2) this strain is lower in harder tissues than in softer tissues. Therefore, by measuring tissue strain due to compression, tissue stiffness can be estimated. Since malignant breast tissue is generally harder than normal surrounding tissue, tissue hardness observed in elastography becomes the more precise clinical information than manual palpation. The use of quantitative elastography achieves the improvement in breast cancer diagnostic accuracy.

1.1 Introduction

Breast cancer is one of the most common cancers in women around the world. Breast cancer in females is one of the top five cancers that cause high mortality rate in thirteen of the fifteen Asian countries (The Burden of Cancer in Asia 2008). To reduce the cancer-related mortality rate, early detection and treatment are

Y. M. Myint (✉) · M. I. Mohamad Salim · N. P. Utama
Faculty of Biosciences and Medical Engineering, Universiti Teknologi Malaysia, Level 3, V01, Block A, Satellite Building, Skudai-Johor, Malaysia
e-mail: yinmontt@biomedical.utm.my

M. I. Mohamad Salim
e-mail: maheza@biomedical.utm.my

N. P. Utama
e-mail: utama@biomedical.utm.my

K. W. Lai
Biomedical Engineering Department, University Malaya, Kuala Lumpur, Malaysia
e-mail: lai.khinwee@um.edu.my

Y. C. Hum
MIMOS Berhad, Technology Park Malaysia, 57000 Kuala Lumpur, Malaysia
e-mail: yc.hum@mimos.my

K. W. Lai et al., *Advances in Medical Diagnostic Technology*,
Lecture Notes in Bioengineering, DOI: 10.1007/978-981-4585-72-9_1,
© Springer Science+Business Media Singapore 2014

essential. To diagnose the pathological changes in breast tissue, palpation is widely used as a screening procedure. Palpation of a mass relies on the fact that tumors most often are harder or stiffer than the tissue in which they are embedded (Garra 2011). Pathological changes in tissue are normally interrelated with elasticity changes. The description of the hardness of pathological tissue according to the cancer disease can be found in medical literature. Scirrhous carcinoma of the breast, which is the most common cancer of the breast and constitutes about three-fourth of all breast cancers, has been described as stony hard, but other types of breast cancers (e.g., intraductal and papillary carcinoma) are soft (Ariel and Cleary 1987; Cespedes et al. 1993). In many cases, although the stiffness of the tissue is different from the surrounding tissues, because of its small size and/or its deep location in the body, it is very difficult to detect by manual palpation. As well, tumors of the breast can be invisible or barely visible in standard ultrasound examinations in early stages (Ophir et al. 1999; Garra et al. 1997).

Over the time of two decades, the mechanical properties of tissue system are investigated. The elastic properties of the soft tissues depend on their molecular building blocks and on the microscopic and macroscopic structural organization of these blocks (Fung 1981). The biological tissue systems have been idealized as homogeneous, isotropic elastic materials.

Different modes of propagation of elastic waves in homogeneous media are determined primarily by their bulk (K) and shear (G) elastic moduli. In biological soft tissue, K is very much greater than G. The bulk properties are determined by the molecular composition of the tissue, and the shear properties are determined by the higher level of tissue organization (Ophir et al. 1997; Sarvazyan et al. 1991). Since deformable soft tissues are essentially volume incompressible (i.e., their Poisson's ratio $v \sim 0.5$), their shear moduli are proportional to their longitudinal (Young's) moduli (Ophir et al. 1997; Saada 1983). Therefore, estimation and imaging of the Young's moduli of tissue should principally convey information about shear properties and hence the higher level of tissue organization. Tissue is anisotropic, and its strain–stress relationship is nonlinear. Generally, an elastic material is defined as the material that has linear relationship between stress and strain. So, tissue is inelastic and exhibits viscoelastic properties such as hysteresis, stress relaxation, and creep (Fung 1981).

The stiffness parameter of the tissue is related to the elastic modulus and its geometry. It cannot be measured directly. Some kind of mechanical stimulus must be propagated into the tissue, and the resulting internal tissue motions must be detected by the techniques such as ultrasound, MRI, or other diagnostic imaging modalities that can track minute tissue motion with high precision. The ultrasonic imaging of tissue elasticity or stiffness parameters has been concentrated in recent researches. Ultrasonic elastography is an imaging modality capable of detecting the stiffness of biological tissues by producing an image based on the strain wave propagating in the interest tissue. It provides non-invasive assessment of variations in tissue elasticity. Tissue elasticity can change with disease, and the shear modulus (or modulus of rigidity) varies over a wide range, differentiating various pathological states of tissues (Seo and Woo 2013). Ultrasonic tissue elasticity imaging methods can be divided into two main groups: (1) Methods where a quasi-static compression is applied to the tissue, and the resulting components of the strain tensor are estimated, and (2) methods where a low-frequency vibration (<1 kHz) is applied to the tissue, and the resulting tissue behavior is inspected by ultrasonic or audible acoustic means (Ophir et al. 1999, 2002; Krouskop et al. 1987). Elastography, one of the quasi-static estimation of tissue strain, estimates the tissue strain using correlation algorithm and elasticity imaging technique estimates tissue strain using signal-phase information. However, in both of these techniques, the local tissue displacements are estimated from the time delays between gated pre-compression and post-compression echo signals. The local strain is estimated from the axial gradient of these signals (Ophir et al. 1999). In sonoelasticity imaging, Doppler method is applied to detect the vibration amplitude pattern of the shear waves in the interested tissue (Lerner and Parker 1987; Lerner et al. 1990). The amplitude and phase of low-frequency wave propagation in the tissue can be used to derive the velocity and depression properties of the wave propagation (Yamakoshi et al. 1990). To measure the tissue flow at interesting points under external vibration, a single element pulsed Doppler instrument is used in Krouskop et al. (1987).

In elastography, the ultrasound transducer is used for tissue compression. In order to minimize the distortion of the signal due to compression, imparted compression is typically about 1 % or less of the total depth of the investigated tissue. Such small compressions thus facilitate accurate time delay analysis. From the combination of the local longitudinal strain information and local longitudinal stress information in the tissue, quantitative images or elastograms of local estimates of the elastic modulus can be generated. These local estimates are displayed in the form of an 8-bit grayscale image. Soft tissues are represented by white, harder tissues are represented by black, and intermediate elasticities are represented by the gray levels. Although elastography is an ultrasonic technique, the elastograms display elasticity information, not echo amplitude information as do conventional sonograms. The appearance of elastograms is therefore quite different from sonograms (Cespedes et al. 1993).

1.2 Fundamentals of Elasticity Theory

The elasticity of a material describes its tendency to resume its original size and shape after being subjected to a compression force. Fluids possess only volume elasticity since they resist only a change in volume. Since solids resist changes in shape and volume, they possess rigidity or shear elasticity and volume elasticity. Strain is defined as the change in size or shape, and it is expressed as a ratio, e.g., the change in length per unit length. The force acting on unit area to produce the strain is known as stress (Wells and Liang 2011).

For a homogeneous isotropic solid, the ratio of stress/strain is called the modulus of elasticity (Nm^{-2} or Pa). There are three moduli used to define the elasticity of a solid.

Table 1.1 Young's modulus
values for human tissues

Tissue	E (kPa)
Artery	700–3,000
Cartilage	790
Tendon	800
Healthy soft tissues (Breast, kidney, liver, prostate)	0.5–70
Cancer in soft tissues (Breast, kidney, liver, prostate)	20–560

- Young's modulus (longitudinal elasticity),
- Shear or torsion modulus (rigidity), G.
- Bulk or volume modulus (volume elasticity), K.

Hooke's law states that strain is proportional to stress. In a one-dimensional simple model of an elastic material,

$$\sigma = E \times \varepsilon \tag{1.1}$$

where E is Young's modulus, σ is axial strain, and ε is axial stress. Young's modulus can be used to predict compression as a result of axial stress. Typical values of Young's modulus for human tissues are given in Table 1.1 (Hoskins et al. 2010). When a material is stressed, its breadth contracts and its length extends. Poisson's ratio, v, is defined from this contraction and extension. It is given by

$$v = \frac{\text{lateral contraction per unit breadth}}{\text{longitudinal extension per unit length}} \tag{1.2}$$

and

$$G = \frac{E}{2(1+v)}, \tag{1.3}$$

$$v = \frac{E}{2G} - 1, \tag{1.4}$$

$$K = \frac{E}{3(1-2v)}. \tag{1.5}$$

For a triaxial stress, the total normal strains are given by

$$\varepsilon_x = \frac{\sigma_x}{E} - \frac{v\sigma_y}{E} - \frac{v\sigma_z}{E} = \frac{1}{E}\left[\sigma_x - v\left(\sigma_y + \sigma_z\right)\right]$$

$$\varepsilon_y = \frac{\sigma_y}{E} - \frac{v\sigma_x}{E} - \frac{v\sigma_z}{E} = \frac{1}{E}\left[\sigma_y - v\left(\sigma_x + \sigma_z\right)\right]$$

$$\varepsilon_z = \frac{\sigma_z}{E} - \frac{v\sigma_x}{E} - \frac{v\sigma_y}{E} = \frac{1}{E}\left[\sigma_z - v\left(\sigma_x + \sigma_y\right)\right]$$

Alternatively, in matrix form,

$$
\begin{bmatrix} \varepsilon_x \\ \varepsilon_y \\ \varepsilon_z \end{bmatrix} = \frac{1}{E} \begin{bmatrix} 1 & -v & -v \\ -v & 1 & -v \\ -v & -v & 1 \end{bmatrix} \begin{bmatrix} \sigma_x \\ \sigma_y \\ \sigma_z \end{bmatrix}. \tag{1.6}
$$

By solving for the direct stresses it is obtained;

$$
\begin{bmatrix} \sigma_x \\ \sigma_y \\ \sigma_z \end{bmatrix} = \frac{vE}{(1-2v)(1+v)} \begin{bmatrix} \frac{1-v}{v} & 1 & 1 \\ 1 & \frac{1-v}{v} & 1 \\ 1 & 1 & \frac{1-v}{v} \end{bmatrix} \begin{bmatrix} \varepsilon_x \\ \varepsilon_y \\ \varepsilon_z \end{bmatrix}. \tag{1.7}
$$

And the shear strain is given by

$$
\begin{bmatrix} \tau_{xy} \\ \tau_{xz} \\ \tau_{yz} \end{bmatrix} = G \begin{bmatrix} \gamma_{xy} \\ \gamma_{xz} \\ \gamma_{yz} \end{bmatrix} \tag{1.8}
$$

where τ is shear stress, and γ is shear strain. The elastic constants, E, v, and G, are related so that there are only two independent constants. The speeds at which mechanical waves propagate in a solid are given by the following equations (Postema 2011).

$$
\text{the longitudinal wave speed, } c_1 = \left(\frac{K}{\rho}\right)^{1/2} \tag{1.9}
$$

$$
\text{the shear wave speed, } c_s = \left(\frac{G}{\rho}\right)^{1/2} \tag{1.10}
$$

where ρ is the tissue mass density.

1.3 Basic Stiffness Data of Breast Tissue

The ability of a tissue to deform its shape when a mechanical force is applied and to recover its original shape after removing the force is referred to as the tissue elasticity. Tumor tissue is harder than normal tissue. In general, soft tissues are anisotropic, viscoelastic, and nonlinear. However, they are usually assumed as linear, elastic, and isotropic for the purpose of analytical simplification (Ophir et al. 2002; Gao et al. 1996). Soft tissues comprise both solid and fluid components, and so their mechanical properties are being somewhere between those of the materials. The ratio G/K of soft tissues is close to a few tenths for solid materials, where as that of liquid is zero. Poisson's ratio for soft tissues is usually between 0.490 and 0.499 (Wells and Liang 2011), which make them mechanically alike to liquid. For most of the solids, the Poisson's ratios are between 0.2 and 0.4 (Sarvazyan et al. 1995).

Table 1.2 Stiffness measurements (kPa) of normal and abnormal breast tissues in vitro

Breast tissue type	5 % pre-compression loading frequency			20 % pre-compression loading frequency		
	0.1	1	4	0.1	1	4
Normal fat ($n = 40$)	18 ± 7	19 ± 7	22 ± 9	20 ± 8	20 ± 6	24 ± 6
Normal glandular ($n = 31$)	28 ± 14	33 ± 11	35 ± 14	48 ± 15	57 ± 19	66 ± 17
Fibrous ($n = 21$)	96 ± 34	107 ± 32	118 vs. 83	220 ± 88	233 ± 59	244 ± 85
Ductal CA ($n = 23$)	22 ± 8	25 ± 4	26 ± 5	291 ± 67	301 ± 58	307 ± 78
Infiltrating ductal CA ($n = 32$)	106 ± 32	93 ± 33	112 ± 43	558 ± 180	490 ± 112	460 ± 178

The values of Young's modulus are usually reported, rather than those of shear modulus in the literature. This is because tissue is almost incompressible. The published data for Young's moduli of breast tissues is listed in Table 1.2 (Krouskop et al. 1998). Most cancers feel stiffer on palpation because they have a lower strain value and a higher Young's modulus. Breast tumors can have a much higher Young's modulus than the surrounding normal tissue, they can be 4 to 10 or more times stiffer, and this ratio is called the relative Young's modulus or the contrast.

The elastic moduli of the different breast tissues did not change with the frequency of the applied displacement over the experimental range of the strain rates (Krouskop et al. 1998). It can be clearly shown in Table 1.2 that the breast fibrous tissues are stiffer than glandular tissues and they are in turn harder than adipose tissue. Tumors with the infiltrating ductal carcinomas are apparently stiffer than the tumors with ductal carcinomas.

1.4 Elastography

It is a direct imaging technique of the strain and Young's modulus of tissues (Ophir et al. 1999; Gao et al. 1996; Sarvazyan 1993). The images produced by elastography are high-resolution images and called elastograms. Elastograms display the axial or lateral strains. The basic principal of elastography method will be described here.

In order to avoid the problems due to reflections, standing waves, and mode patterns, quasi-static stress is applied to the tissue. All points in the tissue experience a certain level of three-dimensional strains. Since a rapid compression is applied, only the elastic property is observed and slow viscous properties are neglected although tissues have viscoelastic properties. In elastography, quasi-static uniaxial stress is applied and the estimation of strain along the ultrasound beam (longitudinal strain) is considered in small tissue elements. If one or more of the tissue elements has different stiffness with the others, the level of strain in that element will be higher or lower. And if the tissue element is harder, then it will experience less strain than the softer one.

The longitudinal axial strain is estimated in one dimension from the analysis of ultrasound signals obtained from the medical ultrasound machine. To accomplish this estimation, the tissue is compressed with the ultrasonic transducer along the ultrasonic radiation axis by a small amount, and then the first set of digitized radio frequency (RF) echo lines reflected from the interesting tissue region is acquired, and then the second, post-compression set of echo lines from the same region of interest is acquired. Congruent echo lines are then subdivided into small temporal windows, which are compared pairwise by using cross-correlation techniques (Foster et al. 1990). From this comparison, the change in arrival time before and after compression can be estimated. Because of the small magnitude of applied stress, small distortions of the echo lines occur and the arrival time changes are also small. The local longitudinal strain is estimated as in Eq. 1.11 (Ophir et al. 1991).

$$e_{11.\text{local}} = \frac{(t_{1b} - t_{1a}) - (t_{2b} - t_{2a})}{(t_{1b} - t_{1a})} \tag{1.11}$$

where t_{1a} = arrival time of pre-compression echo from the proximal window, t_{1b} = arrival time of pre-compression echo from the distal window, t_{2a} = arrival time of post-compression echo from the proximal window, and t_{2b} = arrival time of post-compression echo from the distal window.

Since elastography is a method that is used to generate the new types of images, properties of elastograms are quite different from the properties of sonogram. Sonograms display the information related to the local acoustic backscatter energy from tissue components, and elastograms display the information related to its local strains, Young's modulus or Poisson's ratio. Figure 1.1 illustrates the general process of creating strain and modulus elastograms (Ophir et al. 1997, 1999). The input to elastography system is the tissue modulus distribution. The output can be either elastogram (strain image) or the modulus image. By applying quasi-static stress compression and restricting by mechanical boundary conditions, the tissue strain is obtained and this is measured with ultrasound system. The strain filter (SF) here is the selective filtering of the tissue strain with the contribution of the ultrasound system parameters (acoustical parameters) and signal processing parameters. The SF predicts the elastogram quality by specifying the elastographic signal-to-noise ratio (SNRe), sensitivity and the strain dynamic range at a given resolution (Varghese and Ophir 1997). The contrast-transfer efficiency (CTE) of the strain elastogram from SF is improved in recursive inverse problem solution block.

1.5 Constant-Transfer Efficiency

Using ultrasonic techniques, some of the local longitudinal components of the strain tensor in the tissue can be determined. The strain elastogram represents only the distribution of tissue elastic moduli, and the local components of the stress

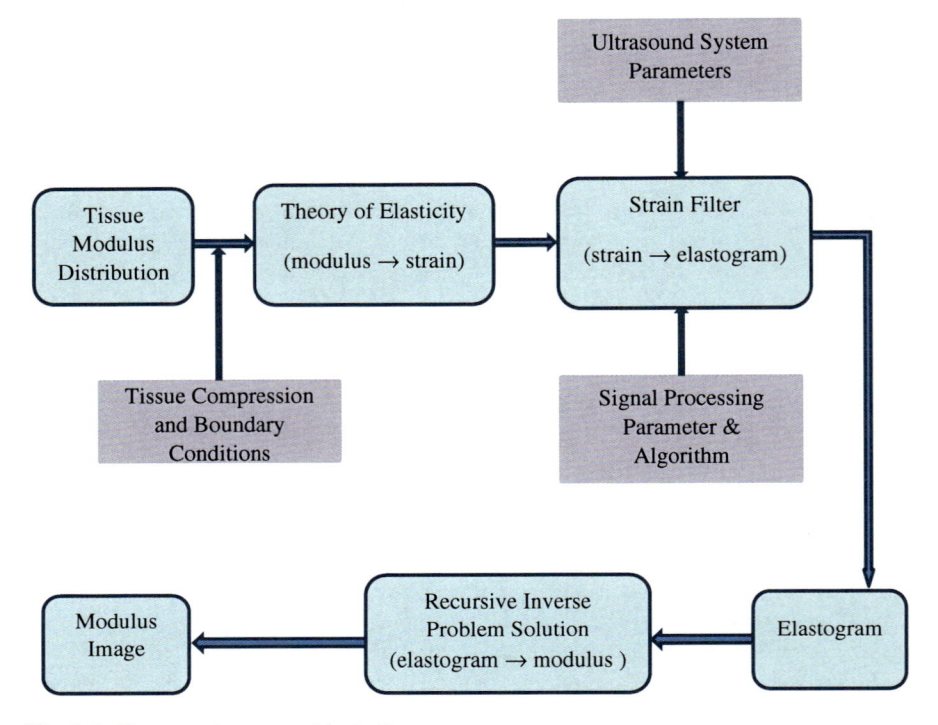

Fig. 1.1 Elastography system block diagram

tensor keep on as unknown (Ophir et al. 1999, 2002). The CTE is defined as the ratio of the observed (axial) strain contrast, C_0, measured from the strain elastogram, and the underlying true modulus contrast, C_T, using a plane-strain-state model. By expressing in decibels, the ratio becomes a difference as

$$\text{CTE(dB)} = |C_0(\text{dB})| - |C_t(\text{dB})| \tag{1.12}$$

$$C_t = \frac{E_t}{E_M}$$

$$\frac{1}{C_0} = \left[\frac{(1-2v)}{C_t + (1-2v)} + \frac{2}{1 + C_t(3-4v)}\right]$$

where
E_t target Young's modulus
E_M background Young's modulus
v Poisson's ratio.

CTE is normalized to the 0 dB, i.e., the maximum efficiency is reached at 0 dB for both hard and soft inclusions.

1.6 Time Delay Estimation in Elastography

Time delays between pre-compression and post-compression echo signals are used to estimate the local tissue displacements, and these are in turn used to estimate the tissue strain. Peak of cross-correlation function between pre-compression and post-compression echo signals are applied to predict the time delays. The quality of the elastography is mainly determined from the quality of TDE. The random noises and the compression force needed for tissue strains introduce the errors in TDE. Then degradation of TDE is being occurred and that degrades the strain estimation.

In strain estimation and imaging, echo signal decorrelation is one of the major artifacts. Decorrelation errors are affected of the relative displacement of scatterers in three dimensions due to tissue compression. For small strains, post-compression echo signal stretching can be totally compensated for signal decorrelation. When the post-compression echo signal is stretched, all the scatters restore within the correlation window. Global uniform stretching could significantly improve the SNRe and expand the strain dynamic range in elastograms (Alam and Ophir 1997). In low-contrast targets and/or low strains, global uniform stretching produces quality elastograms. However, in high-contrast targets, significant over-stretching occurs in the areas of low strains, which causes noteworthy degradation of elastograms in these areas. For these situations, an adaptive axial stretching introduced in Alam et al. (1998a) becomes essential. But axial stretching can recover most of the decorrelation due to scatterer motion in the axial direction and cannot recover the decorrelation due to lateral and elevation motions. The decorrelation due to lateral motion may be compensated using a signal interpolation technique and large improvements in elastographic image quality is obvious (Konofagou and Ophir 1998). During post-compression signal stretching, decorrelation due to the unintended stretching of the transducer point-spread function (PSF) is occurred and the deconvolution filter is used to reduce this artifact (Alam et al. 1998b).

1.7 Strain Filter

A theoretical framework, which characterizes the elastographic system, referred to as the SF was proposed in Ophir et al. (1999) and Varghese and Ophir (1997). By specifying the SNRe, sensitivity, and the strain dynamic range at a given resolution, the SF predicts the elastogram quality. The filtering process is done in strain domain, and the qualified elastograms of a limited range is obtained. The transfer function of SF illustrates the relationship between actual tissue strains and the corresponding strain estimates depicted on the elastogram. Due to the limitations of acoustic parameters from ultrasound system and the signal processing parameters, the range of strain allowed by strain filtering process is limited. Tissue attenuation and speckle decorrelation cause the degradation of SF performance (Ophir et al. 1999). The low strain behavior of SF is determined by the variance,

Fig. 1.2 Theoretical strain filter (Ophir et al. 1999; Varghese and Ophir 1997). *Line with cross symbol* represents the experimental strain response (*ESR*) obtained with global temporal stretching

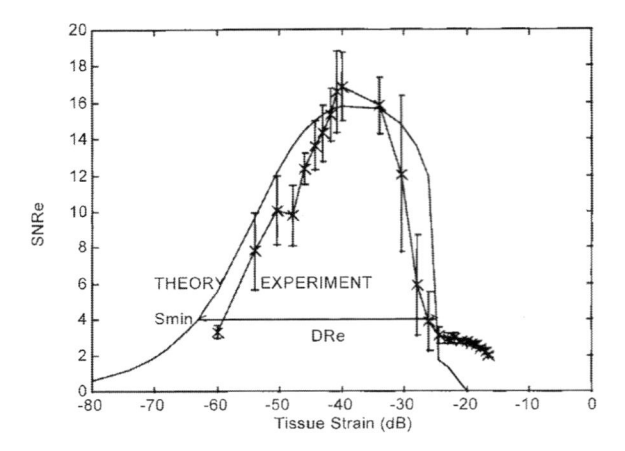

Fig. 1.3 Three-dimensional general appearance of SF. The relation between strain dynamic range and sensitivity, elastographic SNRe and resolution can be shown (Ophir et al. 1999)

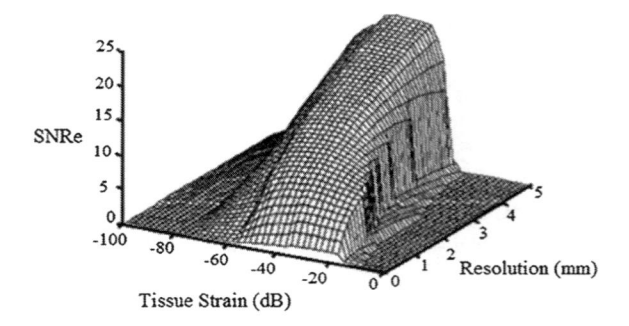

which is computed from the Camer-Rato lower bound (CRLB), and high strain behavior is determined by the rate of decorrelation of congruent signals due to tissue distortion (Walker and Trahey 1995). Figure 1.2 illustrates the theoretical SF (Ophir et al. 1999). The general appearance of the SF is shown in Fig. 1.3. It demonstrates the trade-offs among SNRe and resolution of all strains.

The quantitative measurements of the accuracy and precision of the strain estimation are identified as elastographic signal-to-noise ratio, SNRe. It is defined as

$$\text{SNRe} = \frac{m_s}{\sigma_s} \tag{1.13}$$

where m_s is statistical mean strain estimate, and σ_s is standard deviation for strain noise estimated from elastogram.

The non-stationary pre-compression and post-compression RF echo signals, frequency-dependent attenuation and lateral and evaluation signal decorrelation introduce the non-stationarities in the strain estimation process. To be the more realistic and optimize the SF performance, the non-stationary noise source would be incorporated into the SF formalism. These non-stationary effects on SF is depicted in Fig. 1.4 (Ophir et al. 1999).

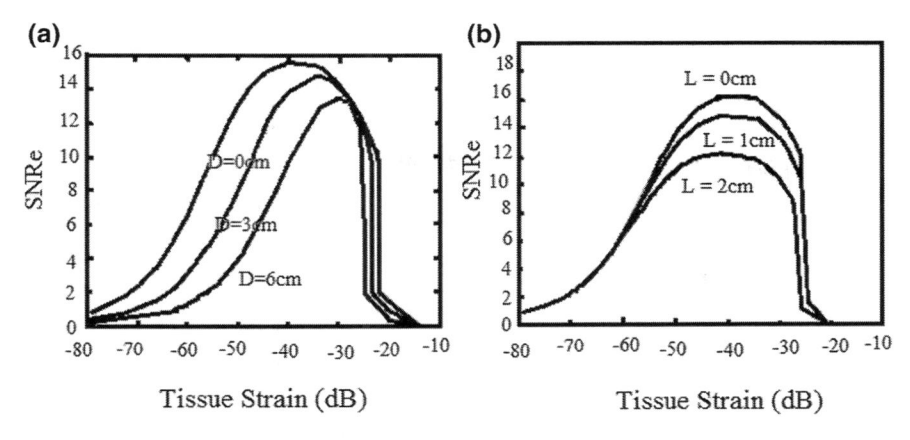

Fig. 1.4 Non-stationary effect on strain filter due to **a** linear frequency-dependent attenuation. **b** Different lateral positions along the transducer aperture and away from the center

Fig. 1.5 Three-dimensional plot of CNRe curves

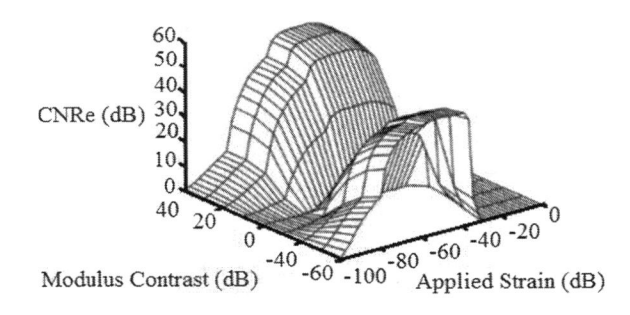

The combination of SF and the elastic contrast properties (CTE) of tissue can predict the elastographic contrast-to-noise ratio (CNRe). It is an important parameter to determine the detectability of lesion in elastograms. An upper bound on the CNRe can be calculated using the Fisher discriminant statistics in Eq. 1.14 (Ophir et al. 1999):

$$CNRe = \frac{2(s_1 - s_2)^2}{\sigma_{s1}^2 + \sigma_{s2}^2} \tag{1.14}$$

where s_1 and s_2 are spatial strain average of target and background, and σ_{s1}^2 and σ_{s2}^2 are the spatial variance of the target and background, respectively. The highest values of CNRe can be obtained when the differences in mean strain values are large or the sums of variances of the strain are small. At low modulus contrast, CNRe is improved due to small strain variances. And at high modulus contrast, CNRe is improved due to large difference of stain means. Three-dimensional plot of CNRe is shown in Fig. 1.5.

In order to get the best estimations of strain, spatial resolution of elastography must be counted. The upper bound on the elastographic axial resolution is proportional to the length of the point-spread function (PSF) of the transducer. That proportionality constant factor is generally between 1 and 2. In practical approach, the resolution highly depends on the window size as long as window size is larger than PSF.

1.8 Strain Elastography

Strain techniques and shear wave techniques are ultrasound elastographic techniques, which are classified based on the underlying measurement principle. Strain techniques rely on the compression of the tissues and resulting tissue deformation and strain using ultrasound. These are noted as 'static' methods (Hoskins et al. 2010). Strain techniques provide information on strains and estimate the elastic modulus. The basic elastic modulus estimation steps are as follows (Hoskins et al. 2010):

Apply known stress: Compressing or stretching the tissues by applying known force or stress.

Measure change in dimensions: Tissue dimension changes are measured, and the strain is calculated with Eq. 1.15.

$$\text{strain} = \frac{\text{changes in length after compression}}{\text{length before compression}} \tag{1.15}$$

Estimate Young's modulus: Young's modulus is estimated from Eq. 1.16.

$$E = \frac{\text{stress}}{\text{strain}} \tag{1.16}$$

Commercial ultrasound systems generally use the transducer itself to compress the tissues. There are two main methods for estimation of tissue movement, one using the amplitude of the received echoes extracted from the RF data and one using the Doppler (DTI) data. Figure 1.6 illustrates a uniform tissue in which there is a stiff lesion. There can be seen the A-scan lines which passing through the center of the lesion pre-compression and post-compression. Figure 1.7b shows the received echoes as a function of depth. As a result of the compression, the tissue movement at each point is predicted. And from this displacement, the strain is estimated (Hoskins et al. 2010).

Estimation of tissue displacement relies on the signal processing methods, essentially, search algorithms. They identify any change in position or stretching of image features. Most of the simple method assumes that the tissue only moves in the direction of the A-line. However, the tissue spreads sideways when it is compressed. So, to improve the estimation of strain and the appearance of the strain image, not only A-line (axial strain) but also between adjacent A-lines (lateral and elevation strains) should be counted to in elastogram creating.

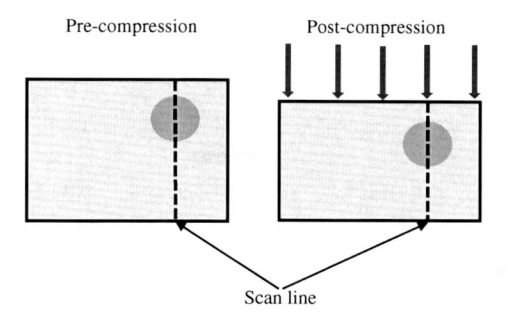

Fig. 1.6 A scan lines on pre-compression and post-compression

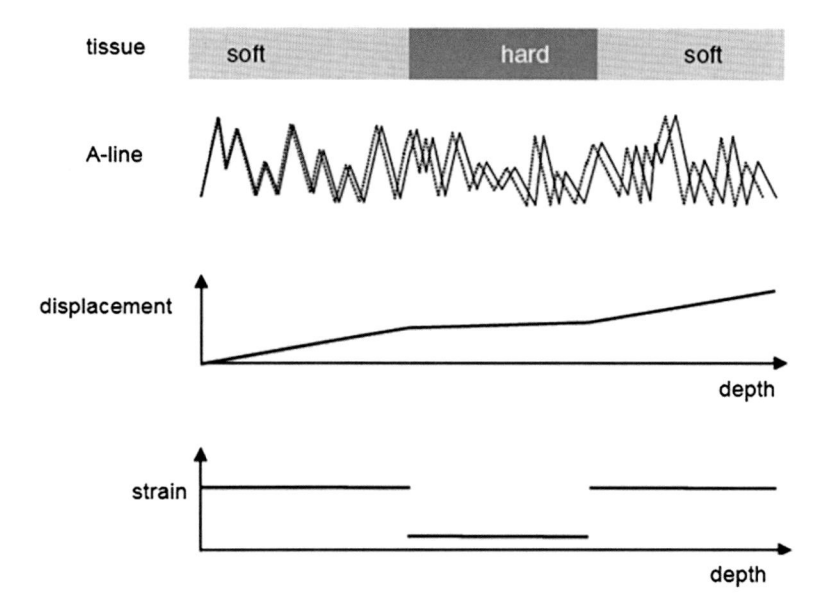

Fig. 1.7 Illustration of estimation steps of strain from A-scan lines pre- and post-compression. **a** Ideal tissue consisting of a hard lesion surrounded by softer tissue; **b** A-lines: displacement between pre-compression and post-compression line (*solid* and *dotted*, respectively); **c** displacement estimated from (**b**); and (**d**) strain estimated from (**c**)

1.8.1 Axial Strain Elastography

As discussed earlier, elastography is a three-dimensional problem according to the tissue motions: axial, lateral, and elevation directions. However, this section discusses elastograms estimated approach concentrating only on axial and lateral and elevation directions are ignored.

Time delay estimation (TDE) is a very important elastographic parameter that determines the quality of elastography as already discussed in previous section. Echo signal decorrelation plays an important role to qualify the strain estimation and imaging. When the post-compression echo signal is stretched, all the scatterers in the correlation window are realigned and those remove the mean intra-window strain. Global stretching algorithm can improve the SNRe and expand the strain dynamic range in elastograms. So, before estimate the tissue displacement, a uniform global stretching of post-compression echo is highly preferred. Although global stretching is advisable techniques for low-contrast target, it is not appropriate for high-contrast targets. For those, adaptive stretching is introduced.

Purpose of temporal stretching in elastography is to advance TDE. But the quality of elastograms depends on the proper selection of stretching methods. However, in turn, the proper temporal stretching factor depends on the local strain. In this case, the local strain is unknown parameter what is to be estimated. Moreover, in elastically inhomogeneous tissue, the strain will vary at different window locations, and thus, the stretching factor will have to be varied. So the iterative algorithm is introduced. In that approach, local temporal stretching factor is adaptively varied until correlation is maximized. Then, local strain is computed from this stretching factor. Since adaptive stretching is accomplished by only intra-window operations, it can remove the noise, which is amplified by gradient operation. Both global and adaptive stretching methods increase the dynamic range of elastography.

In addition, the decorrelation is used to estimate the delay and strain, and correlation coefficient is used to estimate the tissue motion. In freehand elasticity imaging, decorrelation coefficient factor has poor precision as a strain estimator and correlation coefficient has an advantage of simplification. But to use this estimator, the recognition ability and the care are needed. Another estimator to measure the small tissue displacement is phase-band method (Wilson and Robinson 1982). In this estimation approach, some bandpass filtering is needed before phase computation, and it makes the loss in spatial resolution. Least-square strain estimator (LSQSE) is one of the methods proposed as strain estimator (Kallel and Ophir 1997). This method can improve SNRe significantly since it can reduce the amplification of displacement noise due to gradient operation. Increment of elastographic sensitivity can be seen in LSQSE results, but the strain contrast and spatial resolution are reduced. In Alam and Parker (1995), butterfly search technique is developed for complex envelope signals. This technique is based on Schwartz's inequality.

1.8.2 Lateral Strain Elastography

Basically, axial strain components are used to create elastogram, and lateral and elevation components are ignored. However, all three components are necessary to get the actual displacement of tissue. And moreover, the lateral and elevation components become the main cause of interfere to axial components because of their decorrelation factors. So, the techniques to use lateral component in elastogram become interested. The interpolation scheme for lateral component of tissue displacement is stated in Konofagou and Ophir (1998). Method of producing high precision lateral displacements introduced in Konofagou and Ophir (1998) is as follows (Ophir et al. 1999):

- Via a weighted interpolation method, post-compression A-lines are generated.
- Pre-compression segments are cross correlated with original and interpolated post-compression segments.
- The amount of lateral displacement is determined from the location of maximum correlation.
- Lateral strain is derived from lateral displacement with least-square algorithm.

1.8.3 Modulus Elastography

Mechanical artifacts and contrast-transfer efficiency limits the quantitative strain elastography. In the case of relatively simple arrangement of elastic inhomogeneities, the axial strain image alone is appropriate. But for complex conditions, strain image alone is not sufficient (Ophir et al. 1999). So, inverse problem (IP) approach becomes as the interesting research areas among the biomedical field. There is a variety of techniques to solve IP in elastography. One of these techniques proposed in Kallel and Bertrand (1996) is based on the use of linear perturbation method. In this technique, the compression stress and the axial displacement measured are used to reconstruct the modulus distribution. The simple strain images are not sufficient to detect and characterize the tissue stiffness, and the technique to solve this problem is introduced in Sumi et al. (1995). This method uses the equations of the forward problem and so that the tissue elastic modulus becomes unknown parameter, while strain and displacement are known values.

1.8.4 Poisson's Ratio Elastography

By assuming under uniaxial stress, Poisson's ratio for a plane strain is defined as

$$v = -\frac{\varepsilon_l}{\varepsilon_a} \tag{1.17}$$

where ε_l and ε_a are the lateral and axial strains, respectively. Poisson's ratio states the compressible property of the material. If $v = 0.5$, then the material is totally incompressible, and if $v = 0$, then the material is totally compressible. The image that displays the distribution of Poisson's ratio in the tissue is derived based on

Eq. 1.17 and that image is referred to as Poisson's ratio elastogram. These elastograms are useful for detection and imaging of fluid transport in local regions of edema, inflammation, or other hydrate poroelastic tissues. And Poisson's elastograms are independent of geometrical artifacts (Ophir et al. 1999; Mak et al. 1987).

1.8.5 Elastography Breast Imaging

Figure 1.8 (Hoskins 2010) displays typical images of strain in hard and soft lesions in breast phantoms. The displayed strain may be presented in color or in grayscale. Strain elastography based on A-line methods is especially suitable for real-time application. It has two advantages:

- immediate visualization of strain and
- the degree of compression tends to be less than that required for DTI methods (Hoskins 2010).

Figure 1.9 (Hoskins 2010) depicts the elastogram, which is estimated using Doppler tissue imaging (DTI) method. Based on the velocity information at each pixel in the image, it is possible to estimate how much the tissue has been stretched or compressed (strain) and how fast the tissue is stretched or compressed (strain rate) (Hoskins and Criton 2010). The processes including in this method are estimating tissue velocities, estimating velocity gradient and estimating strain. When the ultrasound transducer is pushed, the tissues are compressed, and this acts as a movement of the tissues toward the transducer in the image. As well, when the transducer is withdrawn, the tissue moving away from the transducer and this appears in the ultrasound image. Using these movements, DTI image of velocity is derived as shown in Fig. 1.9b. The strain gradient image is described in Fig. 1.9b. The strain values within the stiff tissue are smaller than the adjacent tissue. For the purpose of to get a sufficiently large velocity, the tissue must be pressed in several mm, which is more than the required force for A-line-based methods. The potential artifacts caused by this requirement can be shown in Fig. 1.9d. One artifact is the displacement of lesions out of the imaging plane, and the other is the halo artifact.

1.9 Shear Wave Elastography

Based on the generation of shear waves and from the measurement of shear wave velocity within the tissues using ultrasound, elastic modulus is estimated in this technique. This is defined as 'dynamic' methods.

When a shear force is applied on a cube as shown in Fig. 1.10, it drags one surface of cube in its direction and transmits through the surrounding. This action causes the distortion or sheared, in the direction of the force. The ability of the

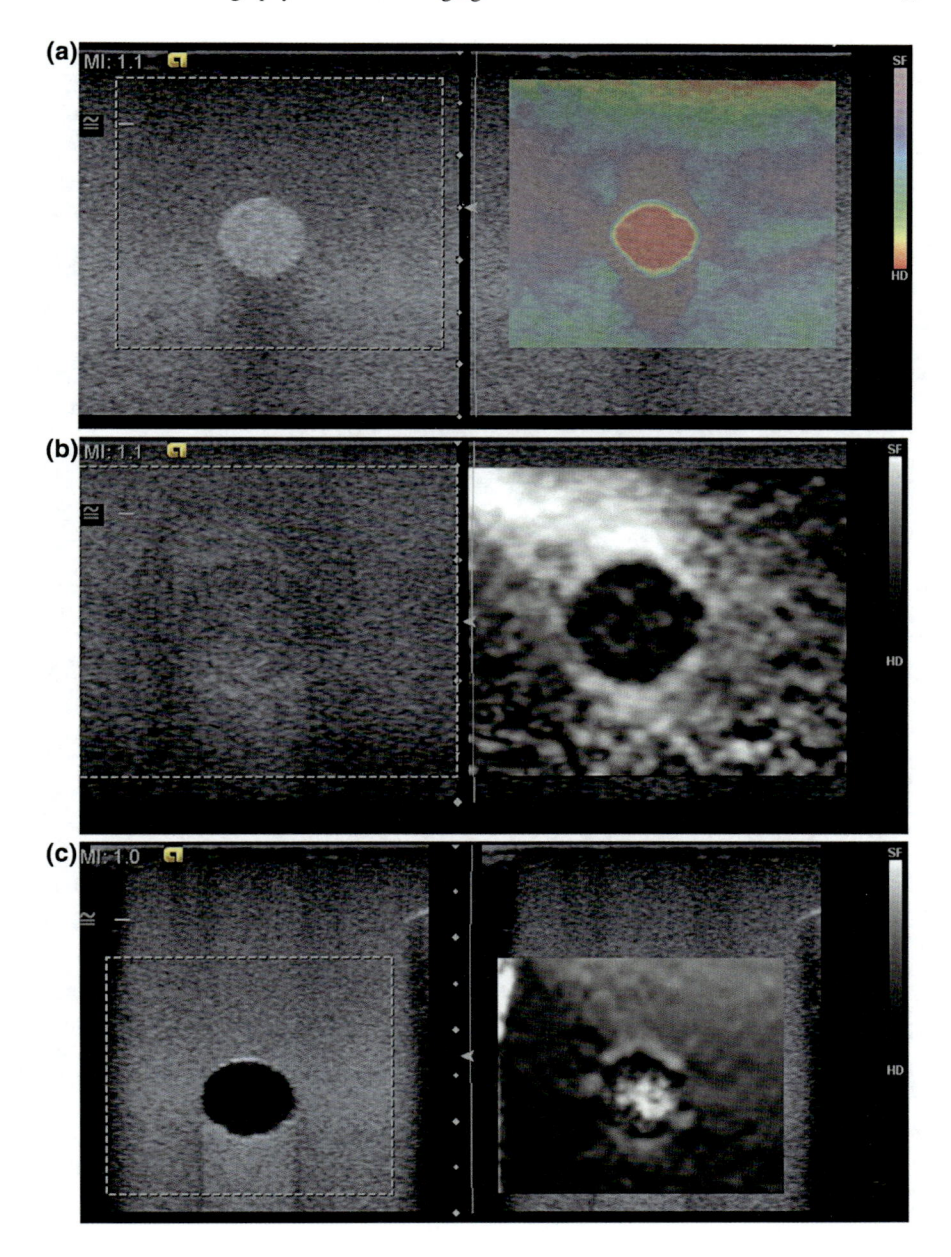

Fig. 1.8 Strain images for lesions in breast phantoms. In each of the figures, the B-mode image is on the *left* and the strain image is on the *right*. **a** Stiff echogenic lesion on B-mode, with low strain (*red*); **b** stiff lesion in which the lesion has the same echogenicity as the surrounding tissue, with low strain (*black*); **c** fluid-filled structure shown as an echo-free region on the B-mode, with high strain (*white*) (Hoskins 2010)

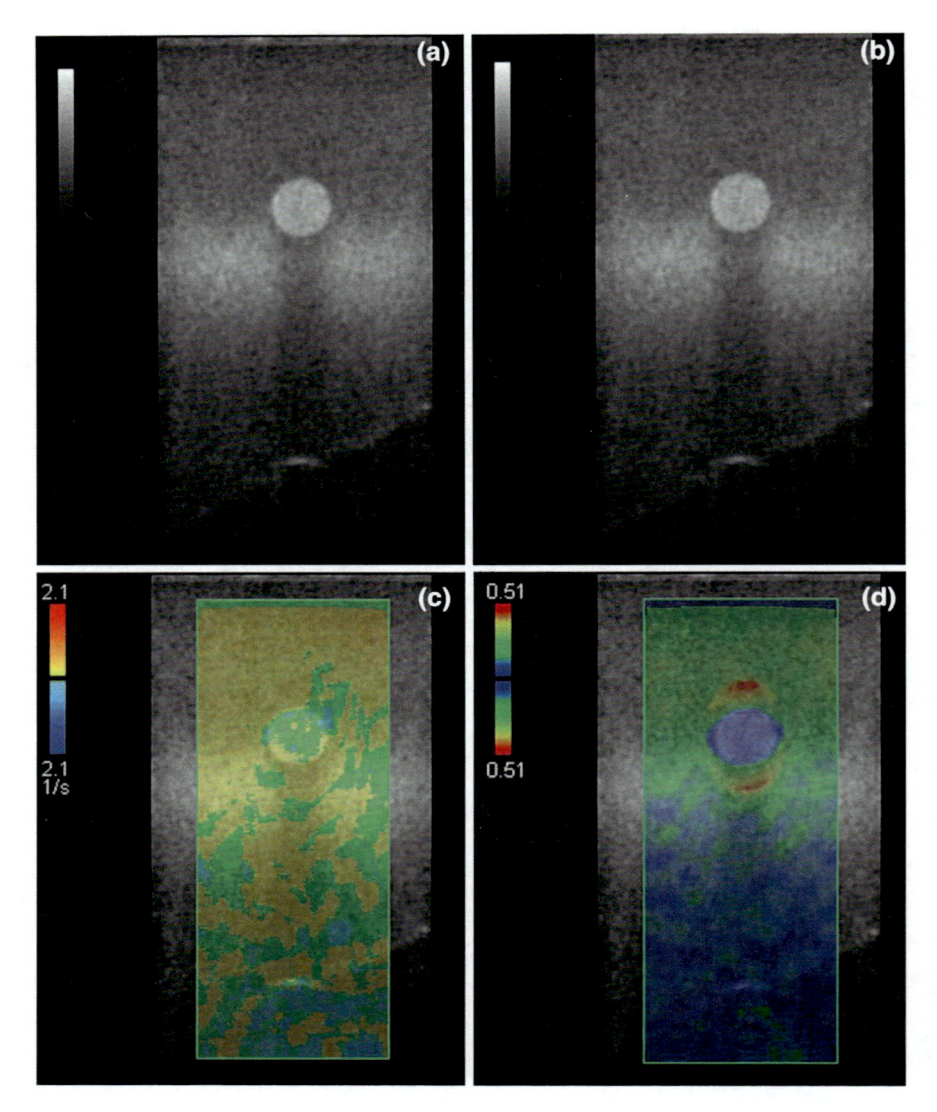

Fig. 1.9 DTI estimation of the strain. **a** B-mode image of a stiff lesion in a breast phantom; **b** DTI velocity image; **c** velocity gradient image; and **d** strain image showing reduced strain in the lesion (Hoskins 2010)

material to withstand a shear force, F, is defined as shear modulus. The amount of shear is represented by an angle, θ. The shear modulus can be described as

$$G = \frac{\text{shear stress}}{\text{shear strain}} = \frac{F/A}{\tan \theta} \tag{1.18}$$

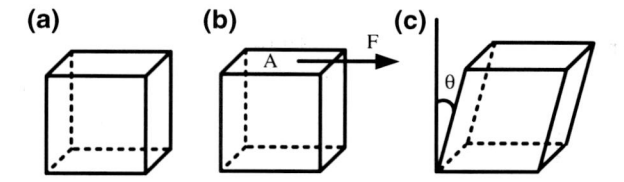

Fig. 1.10 Shear force and distortion. Compression: (**a**) cube of material (**b**) apply a force acting parallel to one of the sides resulting in (**c**) being pulled the cube

Table 1.3 Typical density value of soft tissues	Tissue	Density (kg m^{-3}) mean (range)
	Fat	928 (917–939)
	Muscle–skeletal	1,041 (1,036–1,056)
	Liver	1,050 (1,050–1,070)
	Kidney	1,050
	Pancreas	1,040–1,050
	Spleen	1,054
	Prostate	1,045
	Thyroid	1,050 (1,036–1,066)
	Testes	1,040
	Ovary	1,048
	Tendon (ox)	1,165
	Average soft tissues (excluding fat and tendon) mean (S.D.)	1,047 (5)

And the shear wave speed

$$c_s = \left(\frac{G}{\rho}\right)^{1/2}$$

where c_s is the speed of the wave, G is the shear modulus, and ρ is the density.

By assuming the tissue to be incompressible (no change in density) and uniformly elastic, the shear modulus G is related to Young's modulus E as in Eq. 1.19.

$$E = 3G = 3\rho c_s^2 \tag{1.19}$$

Estimation of tissue stiffness from shear wave velocity is accomplished by the following steps.

Inducing shear waves in the tissue: When the tissue is vibrating, shear waves will be produced, which will propagate in all directions. The propagation speed depends on the local density and the local elastic modulus. Typical frequencies of shear waves are in the range 10–500 Hz. The speed of propagation of shear waves is typically 1–10 ms^{-1}.

Propagation speed measurement: The speed of propagation of the shear waves in the interest tissue region is measured using the ultrasound system.

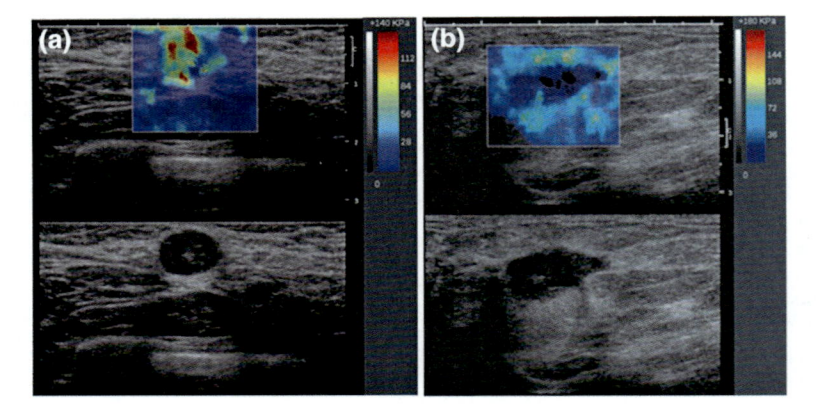

Fig. 1.11 Supersonic shear wave imaging in the breast. **a** B-mode image has high stiffness, confirmed as a metastatic lymph node by pathology. **b** B-mode image has low stiffness, confirmed as an abscess by pathology (Hoskins 2010)

Estimating tissue stiffness: The stiffness value is calculated by Eq. 1.19. As density cannot be measured in vivo non-invasively, it is needed to assume the density of tissues. Values for the density of various tissues are provided in Table 1.3 (Hoskins 2010). The average density for a range of soft tissues (breast, prostate, liver, and kidney) is $1,047 \pm 5$ kg m^{-3}. Figure 1.11 shows the example application of supersonic shear wave imaging in the breast.

References

Alam SK, Ophir J (1997) Reduction of signal decorrelation from mechanical compression of tissues by temporal stretching: applications to elastography. Ultrasound Med Biol 23:95–105

Alam SK, Parker KJ (1995) The butterfly search technique for estimation of blood velocity. Ultrasound Med Biol 21:657–670

Alam SK, Ophir J, Konofagou E (1998a) An adaptive strain estimator for elastography. IEEE Trans UFFC 45:461–472

Alam SK, Ophir J, Cespedes I (1998b) A deconvolution filter for improvement of time-delay estimation in elastography. IEEE Trans UFFC 45(6):1565–1572

Ariel IM, Cleary JB (1987) Breast cancer diagnosis and treatment. McGraw-Hill, New York

Cespedes I, Ophir J, Ponnekanti H, Maklad N (1993) Elastography: elasticity imaging using ultrasound with application to muscle and breast in vivo. Ultrason Imaging 15:73–88

Foster GC, Embree MP, O'Brien WD (1990) Flow velocity profile via time-domain correlation: error analysis and computer simulation. IEEE Trans Ultrason Ferroelec Freq Control 37:164–174

Fung YC (1981) Biomechanical properties of living tissues, Chap. 7. Springer, New York

Gao L, Parker KJ, Lerner RM, Levinson SF (1996) Imaging of the elastic properties of tissue—a review. Ultrasound Med Biol 22:959–977

Garra BS (2011) Tissue elasticity imaging using ultrasound. Appl Radiol 40:24

Garra BS, Cespedes EI, Ophir J, Spratt RS, Zuurbier RA, Magnant CM, Penananen MF (1997) Elastography of breast lesions: initial clinical results. Radiology 202:79–86

Hoskins P (2010) Elastography, Diagnostic ultrasound physics and equipment, Chap. 15, 2nd edn. Cambridge University Press, Cambridge, pp 196–214

Hoskins P, Criton A (2010) Colour flow and tissue imaging. In: Diagnostic ultrasound physics and equipment, 2nd edn. Cambridge University Press, Cambridge, pp 121–141

Hoskins PR, Martin K, Thrush A (eds) (2010) Diagnostic ultrasound physics and equipment, 2nd edn. Cambridge University Press, Cambridge

Kallel F, Bertrand M (1996) Tissue elasticity reconstruction using linear perturbation method. IEEE Trans Med Imaging 15:299–313

Kallel F, Ophir J (1997) A least squares estimator for elastography. Ultrason Imaging 19:195–208

Konofagou EE, Ophir J (1998) A new elastographic method for estimation and imaging of lateral displacements, lateral strains, corrected axial strains and Poisson's ratios in tissues. Ultrasound Med Biol 24:1183–1199

Krouskop TA, Vinson S, Goode B (1987) A pulsed Doppler ultrasonic system for making noninvasive measurements of the mechanical properties of soft tissue. J Rehabil Res Dev 24:1–8

Krouskop TA, Wheeler TM, Kallel F, Hall T (1998) The elastic moduli of breast and prostate tissues under compression. Ultrason Imaging 20:151–159

Lerner RM, Parker KJ (1987) Sono-elasticity in ultrasonic tissue characterization and echographic imaging. In: Proceedings of 7th European community workshop, Nijmegen, The Netherlands

Lerner RM, Huang SR, Parker KJWD (1990) 'Sonoelasticity' images derived from ultrasound signals in mechanically vibrated tissues. Ultrasound Med Biol 16:231–239

Mak AFT, Lai VM, Mow VC (1987) Biphasic indentation of articular cartilage. Part I: theoretical analysis. J Biomech 20:703–714

Ophir J, Ce'spedes EI, Ponnekanti H, Yazdi Y, Li X (1991) Elastography: a quantitative method for imaging the elasticity of biological tissues. Ultrason Imaging 13:111–134

Ophir J, Kallel F, Varghese T, Bertrand M, Cespedes I, Ponnekanti H (1997) Elastography: a systems approach. Ultrasound Med Biol 23(1):89–173 Wiley

Ophir J, Alam SK, Garra B, Kallel F, Konofagou E, Krouskop T, Varghese T (1999) Elastography: ultrasonic estimation and imaging of the elastic properties of tissues. Proc Instn Mech Engrs Part H: J Eng Med 213:203–233

Ophir J, Alam SK, Garra B, Kallel F, konofagou E, krouskop T, Merritt CRB, Righett R, Souchon R, Srinivasan S, Varghese T (2002) Elastography: imaging the elastic properties of soft tissues with ultrasound. J Med Ultrasonics 29(Winter):155–171

Postema M (2011) Fundamentals of medical ultrasonics. Spon Press, New York

Saada S (1983) Elasticity, theory and applications. Pergamon Press, New York

Sarvazyan AP (1993) Shear acoustic properties of soft biological tissues in medical diagnosis. J Acoust Soc Am Proc 125th Meet 93:2329

Sarvazyan AP, Skovoroda A, Vucelic D (1991) Utilization of surface acoustic waves and shear acoustic properties for imaging and tissue characterization

Sarvazyan AP, Skovoroda AR, Emelianov SY, Fowlkes JB, Pipe JG, Adler RS, Buxton RB, Carson PL (1995) Biophysical bases of elasticity imaging. Acoust Imaging 21:223-240 (Springer, US)

Seo JK, Woo EJ (2013) Nonlinear inverse problems in imaging. Wiley, London

Sumi C, Suzuki A, Nakayama K (1995) Estimation of shear modulus distribution in soft tissue from strain distribution. IEEE Trans Biomed Eng 42:193–202

The Burden of Cancer in Asia, Pfizer Medical Division, 2008

Varghese T, Ophir J (1997) A theoretical framework for performance characterization of elastography: the strain filter. IEEE Trans Ultrason Ferroelec Freq Control Appl Muscle Breast vivo Ultrason 44:164–172

Walker FW, Trahey EG (1995) A fundamental limit on delay estimation using partially correlated speckle signals. IEEE Trans Ultrason Ferroelec Freq Control 42:301–308

Wells PNT, Liang H-D (2011) Medical ultrasound: imaging of soft tissue strain and elasticity. J R
    Soc Interface 8:1521–1549
Wilson LS, Robinson DE (1982) Ultrasonic measurement of small displacements and
    deformations of tissue. Ultrason Imaging 4:71–82
Yamakoshi Y, Sato J, Sato T (1990) Ultrasonic imaging of internal vibration of soft tissue under
    forced vibration. IEEE Trans Ultrason Ferroelec Freq Control 37:45–53

Chapter 2
Review on Segmentation of Computer-Aided Skeletal Maturity Assessment

Yan Chai Hum, Khin Wee Lai, Nugraha Priya Utama, Maheza Irna Mohamad Salim and Yin Mon Myint

Abstract Bone age assessment (BAA) is an examination of ossification development with the purpose of deducing the skeletal age of children to monitor their skeletal development and predict their future adult height. Conventionally, it is performed by comparing left-hand radiographs to standard atlas by visual inspection; this process is subjective and time-consuming; therefore, the automated inspection system to overcome the drawbacks is established. However, the automated BAA system invariably confronts with problem in segmentation, which is the most crucial procedure in the computer-aided BAA. Inappropriate segmentation methods will produce unwanted noises that will affect the subsequent processes of the system. The current manual or semi-automated segmentation frameworks have impeded the system from becoming truly automated, objective, and efficient. The objective of this thesis is to provide a solution to the mentioned unsolved technical problem in segmentation for automated BAA system. The task is accomplished by first applying the modified histogram equalized module, then undergoing the proposed automated anisotropic diffusion, following by a novel

Y. C. Hum (✉)
MIMOS Berhad, Technology Park Malaysia, 57000 Kuala Lumpur, Malaysia
e-mail: yc.hum@mimos.my

K. W. Lai
Biomedical Engineering Department, University Malaya, Kuala Lumpur, Malaysia
e-mail: lai.khinwee@um.edu.my

N. P. Utama · Y. M. Myint
Faculty of Biosciences and Medical Engineering, Universiti Teknologi Malaysia, Level 3, V01, Block A, Satellite Building, Skudai-Johor, Malaysia
e-mail: utama@biomedical.utm.my

Y. M. Myint
e-mail: yinmontt@biomedical.utm.my

M. I. Mohamad Salim
Universiti Teknologi Malaysia, Level 3, V01, Block A, Satellite Building, Skudai-Johor, Malaysia
e-mail: maheza@biomedical.utm.my

K. W. Lai et al., *Advances in Medical Diagnostic Technology*,
Lecture Notes in Bioengineering, DOI: 10.1007/978-981-4585-72-9_2,
© Springer Science+Business Media Singapore 2014

fuzzy quadruple division scheme to optimize the central segmentation algorithm, and finally, the process ends with an additional quality assurance scheme. The designed segmentation framework works without the need of resources such as training sets and skillful operator. The quantitative and qualitative analysis of the resultant images have both shown that the designed framework is capable of separating the soft tissue and background from the hand bone with relatively high accuracy despite omitting the above-mentioned resources.

2.1 Introduction

Having a distinct definition to physical maturity is itself a problem, what an to accurately measure it with precise quantitative value; traditional stature measurement does not assure common end points and this complicates the measurement of maturity. Therefore, we are not certain about a child maturity growth by his or her chronological age; therefore, stature measurement is unsuitable for maturity measurement. However, there are some defined events that are certain to normal individuals. Those events are considered to be suitable for maturity measurement. Events during puberty throughout adolescence such as eruption of a certain tooth, occurrence of first menstrual period, degree of testicular, and appearance of pubic hair can be used as indicator for the maturity. The maturity is deducible by the events occurrence. For example, a child that has had one particular event occurred is matured than the other has not had the particular event occurred. Moreover, the extent to which the event has developed can be treated as an indication for the maturity development. These events are regarded as developmental 'milestones' to represent the degree along which an individual has travelled in the maturity growing pathway.

The maturity measurement based on the evens sequence does suffer from certain pitfalls. For example, the events are loosely spaced (Tanner 2001) such that it fails to coverage evenly and completely along the developmental age span (Tanner 2001). Therefore, hand and wrist bones started to be utilized for maturity measurement since hand and wrist bones contain enough sequences to cover the development age span. The appearance of certain bones of hand and wrist occur only during fetal stage and some during puberty stage, and these are all the mentioned sequences that can indicate maturity growth (Aicardi et al. 2000). The measurement of maturity by hand bones sequences involve total of twenty bones: radius, ulna, metacarpals, phalanges, and seven carpals bones develop in all stages along the pathway to full maturity. Therefore, analysis of the hand skeletal development enables the maturity to be measured. The mentioned analysis from hand skeletal bones development in deducing the bone maturity age is named 'bone age assessment (BAA)'.

BAA is used clinically to assess the skeletal development especially in children and adolescents (Cao et al. 2000). As mentioned, estimation of maturation age by chronological age is inefficient. Skeletal maturity or bone age is regarded as a diagnosis indication of growth disorders and also it can be used to predict future adult height of the child (Martin et al. 2011). Conventionally, the left-hand radiograph is utilized to measure skeletal maturation because it is proven that the skeletal growth of hand represents the biological maturity. Such analysis of maturation is predicated on important growth features such as ossification area and calcium position in the ossification area (Roche and French 1970). Diseases such as endocrine disorders (Chemaitilly and Sklar 2010), chromosomal disorders, and early sexual maturation can be indicated through the calculated difference between the skeletal age and biological age (Heinrich 1986).

Currently, there are two types of bone age evaluation systems (Peloschek et al. 2009). First one is regarded as the Greulich-Pyle atlas (Tristán-Vega and Arribas 2008) method (Greulich and Pyle 1959) and the second one is regarded as the Tanner-Whitehouse (TW2) methods (Tanner and Whitehouse 1975). For the first method, patient's hand bone radiograph will be compared with the atlas and the conclusion is drawn; the second method is method where all the clues of skeletal growth are gathered as evidence and then converted into point which then will be used to draw conclusion. The second method TW2 system has been improved into TW3 system (Tanner 2001) in which TW2 gathers the maturity evidences from twenty bones score consisting of the combination of radius, ulna, and short bones (RUS) and carpals whereas TW3 gathers maturity evidences separately from RUS or Carpal scores because research shows that conclusion drawn from the combination of RUS and carpals is not performing as good as analyzing them separately in BAA (Aja-Fernández et al. 2004).

Both methods' reliability and accuracy have been controversial since both of them are inspected visually by physicians or radiologists. This visual inspection by human is considered to be overly dependent on the operator knowledge background and perspective; moreover, it is too time-consuming (Acheson et al. 1963; Tanner and Gibbons 1994; Ontell et al. 1996; Martin et al. 2009). Therefore, in recent years, a lot of computerized system of BAA has been developed especially for TW2 method due to its nature that is more suitable for computational execution (Pietka et al. 2001; Hsieh et al. 2007; Thodberg et al. 2009a, b). However, this kind of computerized system is still developing and far from being perfect due to the system instability and the need for manual operator to execute the system (Jonsson 2002).

In this chapter, the aim is to discuss the very first step of this computerized system: the segmentation of hand bone. This segmentation has to be subject to several constraints in order to achieve stability and autonomous property. The purpose of this segmentation is to capture the bones of the hand skeleton and delineate the hand anatomical outline in the context where the system must be autonomous and the computational efficiency must be high enough to execute instantly. In next part, let us first discuss how other previous researcher in this areas has performed in addressing the segmentation problem and to what extent the current conventional approaches can be used to solve the problem.

2.2 The Attempts for the Hand Skeletal Bone Segmentation

The automated BAA system always go through a preprocessing step, namely the segmentation to eliminate the background, noise, soft tissue region since this information contains no pertinent clues in assessing the bone age in the computerized system (Pietka et al. 1993, 2001; Zhang and Huang 1997; Niemeijer et al. 2003; Aja-Fernández et al. 2004; Somkantha et al. 2011a, b). Not only this information provides no clues, it deteriorates the subsequent stages of computerized BAA system that will affect the accuracy. Nonetheless, most of the conventional techniques adopted in this preprocessing stage are not effective in using the computational resources. Most critically, the segmentation step always involves human operation to perform. Moreover, many techniques execute the segmentation after getting the region of interest (ROI) to ease the process of segmentation (Han et al. 2007; Hsieh et al. 2008). However, this accuracy and performance of ROI search is improvable by executing the algorithm after hand bone segmentation from the soft tissue region. As the significant initial step of the system, the output accuracy and practicality of segmentation are critical since the quality of the computerized BAA system output depends heavily on it (Sotoca et al. 2003).

Substantial efforts have been devoted to research on the preprocessing of hand skeletal bone from background and soft tissue region. Most of the works involve the application of threshold setting which is considered ineffective in the hand bone segmentation due to the fact that the soft tissue region contains pixel intensity that similar to spongy bone of the hand skeletal bone (Smyth et al. 1997; Pietka et al. 2001; Aja-Fernández et al. 2004; Buie et al. 2007). Besides, most of the work, after obtaining the region of interest (ROI), implements the active contour model (Kass et al. 1988) that has inherent weaknesses such as high sensitivity toward intensity gradient, high dependency on initiation location and low ability in growing into concavity (Mahmoodi et al. 1997, 2000; Sebastian et al. 2003). Some works implement the statistical analysis to determine the membership of each pixel, whether it is belong to the bone or the soft tissue region (Tristán-Vega and Arribas 2008; Giordano et al. 2010). Some works combine various techniques segmentation in other field into the hand skeletal bone segmentation (Jong-Min and Whoi-Yul 2008; Somkantha et al. 2011a, b). The development of the study has been summarized the following paragraphs:

As early attempt, Michael and Nelson (1989) propose a CAD system for BAA consists of preprocessing, segmentation, and measurement. They preprocess the image using the histogram equalization, and then it is followed by converting the image into binary image and implementing the threshold method of pixel's intensity to remove the background using the model parameters. By using the model parameter, the main drawback is that the problem of overlapping of pixel intensity in bone and background could not be resolved. Moreover, high sensitivity to illumination change and the soft tissue region surrounding the hand bone further deteriorates the result. Manos et al. (1993, 1994) proposed a framework for the automatic hand-wrist segmentation; they have implemented a technique of region growing and region merging after performing the edge detection during the preprocessing. Along this technique, thresholds are used to determine the edge and growing and merging algorithms. Besides, region growing result relies heavily on the performance of edge detection. Lastly, the region merging depends on gray-level similarity size and connectivity that bear a risk of combining the epiphysis sites that are situated around the metaphysis.

A group of well-known researchers for computerized BAA system Pietka et al. (1993) has conducted a number of studies on computer-aided system of hand bone analysis. Segmentation process in their early studies, thresholding, and dilation technique are used for the bones extraction. The algorithm discussed involves dilation that might ruin the result when bones are approaching each other. In the following attempt by Pietka (1994), she has started to extensively focus on the preprocessing procedure on the bone segmentation from the background using windowing technique to compute the local statistical properties followed by finding the centroid from each peak of the histogram of local window. As further attempt, Pietka et al. (2001) conduct a study on image preprocessing and Epiphyseal/Metaphyseal ROI Extraction in BAA automated system. The proposed method is about employing the method of adaptive thresholding. The statistical value of mean and variance of each window is then computed to determine the ROI utilizing the technique of star-shaped median filter and Lee filtering to segment the bone from soft tissue region after obtaining the ROI. However, the method does not address the problem of segmentation with high reliability. The number of peaks found in each local window is uncertain. Errors of computing would occur in some part of the image.

Sharif et al. (1994) have published a paper on bone edge detection: Segmentation of bone employing edge detection base on the intensity by the derivative of Gaussian (Drog) followed by the employment of thresholding technique. The preprocessing technique implemented by Mahmoodi et al. (1997) involves changing the image into binary, and performing the thresholding method using image histogram to obtain the ROI, the subsequent segmentation of epiphysis within the ROI is implemented through the technique of active shape model (ASM). Similarly, the drawbacks of the method are the sensitivity in illumination change and the soft tissue region. The preprocessing method is used in Mahmoodi et al. (1999) for segmentation of bone using deformable models and a hierarchical bone localization scheme. The method background removing process is performed only after obtaining the ROI. Mahmoodi et al. (2000) adopt binary thresholding to acquire the delineation of the hand, followed by location searching of concave–convex; finally, the segmentation is performed by the method of ASMs.

Sebastian et al. (2003) work on segmenting the carpal bones from CT images using deformable models, and the preprocessing combines the strength of all popular segmentation techniques such as active contour models, region growing and the global competition in seeded region growing, and also the local competition in region competition. The result is satisfying, but it involves complicated and heavy computing consumption while computing the partial differential equation. Active contour model (Pietka et al. 2001) has been used in segmenting the bone; the methods (Cao et al. 2000) c-means clustering algorithm, Gibbs random fields, and estimation of the intensity function have been proposed by Pietka et al. They also proposed Gao et al. (2010) segmentation of hand bone during preprocessing using the analysis on histogram. By inspecting the peak of the histogram, the authors identify the soft tissue region and the background.

Hsieh et al. (2007) incorporate adaptive segmentation method with Gibbs random field at the preprocessing stage. Zhang et al. (2007) suggest segmenting the carpal by non-liner filter as preprocessing follows by adaptive image threshold setting, binary image labeling, and small object removal. However, it involves user-specified threshold and Canny edge detection that are not robust in segmentation. Similarly, Somkantha et al. (2011a, b) segment the carpals bone using a combination of vector image model and Canny edge detector. Han et al. (2007) propose to implement watershed transform and gradient vector flow (GVF) to perform the segmentation where the performance of watershed transform and GVF depends heavily on edge gradient strength. Tran Thi My Hue et al. (2011) proposes to implement watershed transform with multistage merging for the segmentation task. Liu et al. (2008) implement only primitive image processing technique such as edge detection and template matching on the preprocessing segmentation. Giordano et al. (2010) perform the segmentation utilizing the derivative difference of Gaussian (DrDog) techniques followed by thresholding using mean and standard deviation. The impracticality of thresholding, edge detectors, and watershed transform will be discussed latter in next chapter.

The utilization of the state-of-the-art technique of deformable model such as ASM and active appearance model (AAM) of hand bone segmentation has gained considerable attentions in recent years (Thodberg 2002; Thodberg and Rosholm 2003; Thodberg et al. 2009a, b). The strength of this method is that it is well founded on statistical learning theory. However, the main drawback of this technique is that it is not yet developed into a fully automated system. The initialization of the technique is to delineate the hand bone shape, and this thus far is accomplished manually. Manual shape delineation is extremely time-consuming especially for ASM and AAM that require a substantial number of training samples before giving enough information to the algorithm about the changes of the hand bone shape. Furthermore, the training samples have to be sufficient to have a comprehensive coverage of hand shape changes and number of bones in different age groups. In other words, this type of segmentation framework is effective only when both human operators and complete training set are available. Thus, the practicality is limited in situation when those resources are limited.

In conclusion, the current existing segmentation methods and frameworks either are involve in threshold settings or are too dependent on certain resources and image features. This indicates that improvement on hand bone segmentation is necessary in order to practically realize the fully automated computerized BAA system. Thus, this research is to explore this improvement aiming to establish a fully automated segmentation framework that is accurate yet remains less dependent on external resources.

2.3 The Problem of Research Problem

The computerized BAA relies heavily on the performance of the segmentation in order to provide the accurate ROI of the hand for further analysis to assess the bone age. Distinct boundaries of anatomical structures of the ossification sites have implications in features localization and analysis which are essential to assure high reliability of the computerized system. Human visual capability, generally, is able to accomplish this visual recognition task effortlessly, but it is not the case for computational visual system due to various obstacles such as the lacking of prior knowledge to reason about the semantics of the object. There are abundant of hand bone segmentation techniques found in the literatures, but very few of which, if ever, is functioning as an effective and yet remains fully automated. The research problem, therefore, is to explore the question: Is there any method that can realize the goal of performing the automated segmentation task that is relatively much more effective than fundamental segmentation techniques but yet is unaffected by constraints such as training sets and human intervention that are invariably pertain to sophisticated techniques?

The factors that associate with the problem are presented as following:

1. Variability in the radiograph image attributes such as deviations in terms of illumination, number of bones, size of bones, and locations of bone. This variability deviates across different input sources and different age groups of the subjects in radiographs. The problem with this variability is that it impairs the performance consistency or precision of segmentation technique or frameworks once the input radiographs are not as expected.
2. The nature of the preprocessing module or the central algorithm in the framework that demands high degree of human cognitive ability such as visual perception or observation on certain patterns of input. This problem is attributable to devoid of prior knowledge of computational algorithm in recognizing pattern that can be easily perceived by human such as the shape of hand bone and the variability of luminance and illumination. As a consequence, most segmentation framework necessitates explicit labors and hence this problem violates automaticity.
3. The inherent bone intensity property in radiograph that stems primarily from the variations in anatomical density of different parts of the hand bone. As a consequence, two adverse properties for segmentation performance take place:

   (a) The overlapping range of pixel intensity for the cancellous bones, the soft tissue regions, and the compact bones. This overlapping intensity range is the main complication that thwarts any global image processing techniques. For instance, the low mineral density in cancellous bone and soft tissue regions leads to similar degree of penetrations and absorptions of X-ray protons resulting in overlapping intensity effect. Most of the intensity-based image processing that postulates on distinct separation of intensity distribution fails to address this problem.

(b) The non-uniformity within the same category of bone such as cancellous bone or the cortical bone. The radiograph intensity is not evenly distributed. This problem stems primarily from the inhomogeneity of mineral density inside the cancellous bone due to porous structure embodying a variety of spaces that are different in size and mineral density. As a consequence, the degree of absorptions is different and hence giving it an irregular spongy textured appearance which in turn results in large variations of pixel intensity within the same structure of bone. This non-uniformity problem ultimately becomes the problem of overlapping range of intensity problem. Hence, additional processing stages containing higher-dimensional input data are in demand to differentiate the pixel into correct labeling to avoid misclassification.

4. The uneven brightness intensity difference between the edge border of compact bone and soft tissue regions and also between soft tissue regions and background further complicates the problem. This is the main problem that deteriorates the performance of edge-based segmentation technique that depends solely on edge information as main input.

5. The existing segmentation methods, as illustrated in next chapter, either are too simplified to have adequate considerations encompassing various aspects of being a comprehensive technique to perform the segmentation tasks or being too dependent on limited resources such as the availability of training hand bone samples over all age ranges, computational complexity, and the knowledge background of operators.

6. Lack of quality assurance process in the hand bone segmentation frameworks to consider possible artifacts. The segmented hand bone at the first stage of computerized BAA system should not proceed to subsequent processing stage before assessing the quality of the segmented hand bone. Most of the existing works do not incorporate such functionality in the designed segmentation frameworks. As a consequence, the subsequent processing stage such as feature analysis for BAA will accept inferior-quality segmented hand bone as input and hence produces final result that is not reliable.

Partitioning hand bone from radiograph background and soft tissue region is the first stage of computerized BAA system; the performance of this stage underpins the success of subsequent procedures of BAA, which in turn affects the final result. Despite being the significant step in computerized BAA, the automated segmentation remains a challenging problem owing to above-mentioned problems. Conventional segmentation methods and the currently developed segmentation framework designed for hand bone segmentation are generally impractical to be implemented attributable to high dependency on various resources. The most problematic of which is the number of user-specified parameters in developing a fully automated segmentation without any human intervention. In addition, to the best of our knowledge after reviewing a vast number of literatures, the comprehensive critical appraisal on the research area of hand bone segmentation can rarely be found, if ever. Therefore, in this research, an automated and practical segmentation framework is to be proposed techniques and implemented to solve the above-mentioned problem after critically evaluating various existing segmentation techniques and frameworks.

2.4 Authors' Proposed Segmentation Framework

The contribution of authors in this problem is to develop an effective segmentation frameworks consisting of several modules that are fully automated and independent from the completeness of training samples and availability of skillful operators. The technical details can be found in Hum (2013). This proposed segmentation framework produces superior segmentations, yet it remains computationally feasible. The performance edges are summarized as follows:

1. Present the critical appraisal of existing segmentation techniques and existing hand bone segmentations frameworks by first elucidating the postulation of each technique based on and the technical details of each technique, followed by illustrating the strengths and weaknesses of which, and finally reasoning their suitability to perform the segmentation task of hand bone. Despite an abundance of segmentation reviews, we could not find any relevant and comprehensive critical evaluations on the aspect of theoretical and technical in the hand bone segmentation for computerized BAA. Hence, this critical appraisal can contribute by serving as the basis for future reference on the research of this field.
2. Extend the comprehensiveness of existing histogram equalization technique by first assessing the current theoretical and technical architecture of existing histogram equalization methods, and then, we contribute our new insight into revolutionize the conventional perception toward the ultimate goal of histogram equalization by proposing the new histogram equalization framework; then, based on the revolutionized insight, we develop a holistic histogram equalization in terms of luminance preservation, contrast, and detail preservation based on the beta function to preprocess the hand bone radiograph serving the purposes of standardizing and equalizing the non-standardized illumination among radiographs that contain high variations in luminance, improving luminance difference across edge borders in radiographs, reducing variations in luminance difference across edge borders among radiographs and most importantly, enhancing the visual perceptual effect of ossification sites for improving the accuracy in ossification localization and BAA.

3. Extend the body of knowledge of anisotropic diffusion by exploiting the potential of being fully autonomous and adaptive to input radiograph instead of being subjectively tuned by operators to solve the problem of non-uniformity and mitigate the undesired effect of overlapping intensity range. Both contributions as following have profound implication for advancing the field of anisotropic diffusion and provide adequate ground for framework that requires autonomous anisotropic diffusion.

   (a) Address the problem of manual diffusion strength by designing an automated diffusion strength scheme based on the diffusion coefficient function of speckle reducing anisotropic diffusion (SRAD) grounding on the well-founded statistical theory of the relation between sample variance and global variance. The main strength is its computationally attractiveness and practical applicability.

   (b) Address the problem of manual scale selection by designing an automated scale selection. The main strength of which compared with limited existing automated scale selection schemes is that it requires no excessive filtered image before determining to halt the diffusion iteration.

4. Transform the manual and rigid adaptive division scheme into an automated adaptive quadruple division scheme that embodies human cognitive ability. This transformation is significant not only in a narrow sense of hand bone segmentation, but most importantly, it is of a generic breakthrough in the field of image segmentation. This implicit modeling of human intuition and prior knowledge solves the problem of high dependency on explicit human resources to operate the algorithm. Furthermore, the scheme itself is a building block or framework for other segmentation algorithm to determine the optimum region size for algorithm implementation.

5. Incorporate quality assurance module in the segmentation framework to evaluate the appropriateness to serve as input for subsequent stages of computerized BAA. The step is important to eliminate over-segmented regions of hand bone and restore under-segmented regions to further improve the quality of segmented hand bone. This concept provides insight about a feedback system of most of the imperfect segmentation framework system should contain a stage that capable of analyzing the current output and patch up the incompleteness accordingly.

6. In conclusion to contributions, this thesis provides the contrary perception to conventional concept that prone to complicating the segmentation algorithm to seek for enhancement in segmentation performance. Instead, the proposed segmentation framework pioneers the insight postulating that combinations of several customized modules are capable of achieving result that tantamount to result achieved by complicated algorithm or algorithm that demands scarce and limited resources. The conception lies in the strategy to identify, target, and analyze the principal adversities that impede the performance of existing methods; then, based on the analyzed result, we contribute by advancing the concept of adapting the input information to the central segmentation algorithm

(in this thesis, it refers to unsupervised clustering). This concept is of contrast to conventional segmentation framework that tends to complicate the fundamental segmentation algorithm in order to adapt the input information. The strength of each module and the idea of concatenating each of them by utilizing the by-product of each module are breaking new ground in the field of automated image processing and pattern recognition. These instructive insights embrace the potential to spark attentions and generate new research grounds that will in turn contribute in other fields of applications.

2.5 Conventional Segmentation Techniques Performance Discussion

This section provides an overview for traditional or conventional segmentation techniques that are usually adopted in applications related to images segmentation to give the reader an overview of the development of segmentation techniques. The fundamental concept of each technique is presented and the pros and cons of each technique are discussed. Besides, we illustrate the unsuitability of traditional segmentation techniques as hand bone segmentation technique in the context of computerized BAA by analyzing the nature of the technique and by implementing the analyzed technique in hand bone segmentation. This evaluation and implementation of previous techniques in hand bone segmentation are crucial motivate the objective and justify the contribution of this thesis. This section ends with the conclusion that a more advanced technique of hand bone segmentation should be derived instead of using the traditional segmentation techniques.

Thresholding is one of the earliest image segmentation techniques, and yet it remains to be the most widely applied segmentation technique attributable to its simplicity and intuitiveness (Sezgin and Sankur 2004). Thresholding segmentation is normally conducted in spatial domain based on the postulation that both object and background are represented by different range of pixel intensity (Gonzalez and Woods 2007). Basically, there are three categories of thresholding: global thresholding, local thresholding, and dynamic thresholding (Bernsen 1986).

Undoubtedly, the simplest method in thresholding techniques to segment an image is through single global thresholding: this technique based on the concept that if object in the image and other object or background are mutually exclusive in terms of intensity range, then it could be separated in different partition using a single or multiple values of pixels intensity (Lee et al. 1990). In the case of single threshold, it can be represented as following:

$$f(x,y) = \begin{cases} g_1 & \text{if} \quad f(x,y) < T \\ g_2 & \text{if} \quad f(x,y) \geq T \end{cases} \tag{2.1}$$

where g_i is group of pixel that represents an object or background; if a pixel value is less than T, which is the threshold value, then it is grouped into g_1; if a pixel value is more than T, then it is grouped into g_2. The $f(x, y)$ is image pixel intensity in 2D grayscale image in coordination (x, y). The concern of the technique is to classify an image into object and background; this type of grouping is called binarization.

The single thresholding depends on the T. This T value determines the intensity range of an object and the intensity range of the image background. For instance, (if the object is brighter than the background) if an pixel intensity value is more than the threshold value, then the pixel will be classified as object; for the pixels which possesses intensity value less than or equal the threshold value, they will be considered as background. This kind of threshold method is considered as 'threshold above'; another type is 'threshold inside' where the object value is in between two threshold values; similarly, another variant is 'threshold outside' where the value in between the two threshold values would be classified as background (Shapiro and Stockman 2001).

The efficiency of thresholding technique in segmentation mainly depends on two factors: first factor is the property of the image intensity distribution of both object and background. Thresholding technique performs most efficiently when the intensity of input image has distinct bi-modal distribution without any overlapping range of intensity for object and background (Liyuan et al. 1997). Overlapping range of intensity occurs often due to uneven illumination. Besides, the nature of the object itself can lead to overlapping range in which some regions within the objects in input image has overlapping range of intensity to background. As mentioned in previous chapter, one of the natures of X-ray hand bone radiograph is its uneven illumination throughout the image as well as its overlapping range of intensity distribution among soft tissue region, trabecular bone, and cortical bone due to the nature of hand bone and uneven background illumination as well.

The reasons for the inferior quality of segmented hand bone by thresholding can be summarized as follows:

1. Assumption that the whole targeted object (which is the hand bone in our case without soft tissue region) contains similar intensity range. This is always not true for hand bone radiograph as within the hand bone, there are regions of trabecular bone and cortical bone which have different bone density and hence are represented by different range of pixel intensity values in digital image.
2. Assumption that the histogram of targeted object and background (black regions and soft tissue regions) is of perfectly separation into two groups of intensity distributions. This is always not true for hand bone that the histogram of hand bone radiograph is not bi-modal distributed. This can be explained from the nature of hand bones that are formed by three classes of regions: bone, soft tissue regions, and background instead of two.
3. Assumption that there is no overlapping of intensity range between background and targeted object. This is always not true for hand bone as the some of the

intensity in soft tissue regions are identical to the regions in trabecular bones. The global thresholding neglects this intensity overlapping problem.

4. Assumption that the illumination is even in input image. This is always not true in for hand bone radiograph that lower region of hand bone radiograph has more intense illumination relative to upper region of the radiograph. The global thresholding neglects this uneven illumination and this affects the segmentation result.

Another critical problem of single global thresholding is the choice of the threshold value to obtain favorable segmentation result (Baradez et al. 2004). In fact, even the 'best' threshold value is selected, the resultant segmented image in the context of hand bone radiograph and in other medical image processing remain inferior. This fact is inevitable due to the nature of global thresholding and the nature of hand bone segmentation: only one threshold. One improvement for this limitation is by adopting multiple global thresholding (Yan et al. 2005). multilevel thresholding classifies the image into multiple classes (>2) (Tsai 1995). The multiple thresholding can represented as follows:

$$f(x,y) = \begin{cases} g_1 & \text{if} \quad f(x,y) > T_1 \\ g_2 & \text{if} \quad T_1 < f(x,y) \leq T_2 \\ \vdots & \qquad \vdots \\ g_{n-1} & \text{if} \quad T_{n-3} \leq f(x,y) \geq T_{n-1} \\ g_n & \text{if} \quad f(x,y) \geq T_{n-1} \end{cases} \tag{2.2}$$

where g_i is group of pixel that represents an object or background. T_i is the threshold values. The $f(x, y)$ is the image pixel intensity in 2D grayscale image in coordination (x, y).

Multiple thresholding might solve the problem arises from the assumption that the input image is of bi-modal type but solve not the problem arises from assumption that the input image is of even illumination. In next subsection, we would review and examine the local/adaptive thresholding that is claimed to be more effective in tackling the problem of uneven illumination.

Adaptive thresholding is segmentation using different thresholds in different sub-images of input image (Zhao et al. 2000). The input image is firstly divided into a number of sub-images; then in each sub-image, suitable threshold is chosen to perform the segmentation, and this process repeats until all sub-images undergo the thresholding segmentation. Adopting different threshold in different region of the input image is proven to be more effective than global thresholding that it is easier to obtain well-separated bi-modal or multiple-modal distributions in the sub-images, and hence, it improves the segmentation result (Shafait et al. 2008). In addition, sub-images are more likely to have uniform illumination implying that as it could resolve the problem that arises from the non-uniform illumination (Huang et al. 2005).

Undoubtedly, it is a fact that adaptive thresholding performs better than global thresholding in tackling the problem of uneven illumination. There are some difficulties in applying the technique effectively in hand bone segmentation due to the problems as follows:

1. The problem arises from making the assumption there is no intensity overlapping between target object and background.
2. The size of each sub-image is difficult to determine. If the size is smaller or larger than it should be, then the result might be even more inferior than using global thresholding.
3. The size of the sub-images is globally set and is fixed throughout the entire image. Some regions need smaller sub-image whereas some regions need larger sub-image in adaptive thresholding to optimize the segmentation and the computational efficiency.
4. The number of thresholds needed in each sub-image is difficult to determine.
5. The computational cost increases in comparison with global thresholding.

The threshold values are difficult to be set manually as the number of sub-images increases (Buie et al. 2007). In global thresholding as well, the threshold value need to be correctly set in order to optimize the result. We afford to set single global threshold using human inspection. However, when we are dealing with multiple thresholding or adaptive thresholding, automated thresholding is more suitable to decrease repetitive threshold setting by human which is subjective and yet time-consuming. In next subsection, we explore and study about the automated threshold value setting techniques which can be applied in both global thresholding and adaptive thresholding. The implementation of multiple thresholding and adaptive thresholding in hand bone segmentation is illustrated in next subsection using automated threshold values selection to demonstrate that the sole implementation of these technique fail to provide good segmented hand bone.

In global thresholding, each pixel is compared with the global threshold; in local thresholding, each pixel in sub-image is compared with each local threshold which is computed from each sub-image; in dynamic thresholding, each pixel is compared with each dynamic threshold which is computed from sliding a kernel over the input image (Shafait et al. 2008). One of the popular dynamic thresholding methods is Niback method (Niblack 1990).

Generally, dynamic thresholding performs better than global thresholding and local thresholding. However, it has similar drawback as local thresholding that we need to determine the kernel size; the threshold has to be selected manually depending on application. Only suitable selection of kernel size and threshold can produce optimum result of segmentation. In addition, dynamic thresholding consumes much more computational resources relative to local thresholding and global thresholding due to its pixel-wise nature. Besides, in performing the neighborhood operations for dynamic thresholding, the padding problem arises when the kernel approaches the image borders where one or more rows or columns of the kernel are placed out of the input image coordinates.

The main technical issue being frequently discussed is the threshold value selection: the decision to determine the threshold value in which the object and the background could be separated as accurate as possible or the decision to select the threshold value so that the object and the background misclassification rate are lowest. The result of thresholding segmentation process depends heavily on this value. An inaccurate or inappropriate setting of this value will produce disastrous result in thresholding segmentation.

For the choice of threshold value, basically, there are two main methods: the manual threshold selection and the automated threshold selection. Manually determined threshold value heavily relies on human visual system. Threshold value is selected using Visual perception to partition the object from the background; the main drawback of this threshold selection is that it involves human subjective perception toward image quality. Besides, the process itself is extremely time-consuming if the operation involves multiple thresholds. Therefore, it is not practical to determine the threshold value of a large number of images. In short, the manually determined value is not effective.

For automated thresholding method, various methods exist: the simplest method is to utilize the image statistics such as mean, median (second quartile), first quartile, and third quartile, to act as threshold value (De Santis and Sinisgalli 1999): this method performs only relatively well in an image free of noises; the reason is that the noise in the image has influenced the statistic of the image. Typically, if the mean of an image used as threshold value, then it can separate a typical image with object brighter than background into two components; however, while noises exist, the noises have altered the nature that the pixels with intensity more than mean are belonged to the object. Besides, this kind of thresholding method assumes that the object and the background are themselves homogenous. In other words, the object is a group of pixels containing similar pixel intensity; the background is a group of pixels with similar intensity. This assumption has serious limitation especially in medical image segmentation where the target objects like organs or bone are not inherently homogenous. Besides using simple aforementioned statistic in input image, there are other methods to choose the threshold value. In next paragraph, we explore and study different types of automated thresholding techniques that have been developed.

Attributable to the limitations of using simple statistics, various more sophisticated types of thresholding methods based on different techniques in determining the threshold value are proposed: one of the methods is the threshold value selection based on histogram: instead of choosing the mean or median of the image as the threshold value to separate the object and the background, the histogram-based thresholding method determines the threshold value based on the histogram shape assuming that there are distinct range for object and background themselves. The value of a valley point is set as threshold.

In image processing, when the histogram of an image is mentioned, typically we mean a histogram of the values of pixel intensity; the graph of the histogram represents the number of pixels in an image at each intensity value of the pixel in the image. If say in an 8-bit grayscale image, there will be 2^8 possible values and it means that the histogram shows the occurrence frequency of each intensity in the image. In other words, it is a representation of the image statistics based on the number of the specific intensity's occurrence.

Histogram analysis is a popular method in automated thresholding (Whatmough 1991). The postulation is that the information obtained from the physical shape of the histogram of the input image signalizes the suitable threshold value in dividing the input image into meaningful regions (Luijendijk 1991). Conventionally, the intensity bin in the valley between peaks is chosen as threshold to reduce the segmentation error rate. Instead of using manual inspections, by only analyzing the shape of the histogram and compute the intensity bin that represents the valley, the relatively good threshold value can be found (Guo and Pandit 1998).

However, the main drawback of this technique is that it depends too heavily on the shape of pixel intensity distribution. Besides, it has no consideration on the pixels location and the pixel surroundings and this leads to the failure in recognizing the semantic of the input image. This method fails when the input image does not have distinctly separated intensity distribution between the foreground and background due to overlapping of intensity as mentioned in last subsection of global thresholding. This category of automatic threshold selection performs thresholding in accordance with the intensity histogram's shape properties. Utilizing basically the histogram's convex hull and curvature, the intervening valley and peaks are identified (Whatmough 1991).

This concept is based on the facts that regions with uniform intensity will produce apparent peaks in the histogram. If only the image has distinct peaks on each objects in the images, then multiple thresholding is always applicable via histogram-based thresholding. The favorable shapes of the histogram for the purpose of segmentation are tall, narrow and contain deep valleys. This method is less influenced by the noise, but it has drawbacks like assuming the pixels intensity range of the object and background has a certain degree of distinction. If the image has no distinct valley point in the histogram, this method would fail to separate the object and the background. The main disadvantage of this histogram-based thresholding method is the difficulties they meet when they have to identify the important peaks or valleys in the image used for segmentation and classification. In next paragraph, we would explore another main automated thresholding based on clustering.

The edge-based segmentations discussed in the previous subsection attempt to perform object boundaries extraction in accordance with the identified meaningful edge pixels. Region-based segmentations, on the contrary, seek to segment an image by classifying image into two sets of pixels: interior and exterior, based on the similarity of selected image features. In this subsection, we explore and study several classic methods belong to this category.

The region-based segmentation is based on the concept that the object to be segmented has common image properties and similarities such as homogenous distribution of pixel intensity, texture, and pattern of pixel intensity that is unique enough to distinguish it from other object (Gonzalez and Woods 2007). The ultimate objective is to partition the image into several regions where each region represents a group of pixels belong to a particular object.

Another popular region method is seeded region growing; this method grows from seeds which can be regions or pixels; then, the seeds expand to accept other unallocated pixel as its region member according to some specified membership function (Kang et al. 2012).

In comparison with deformable model-based segmentation, region-based segmentation is considered relatively fast in terms of computational speed and resources. Besides, it is certain that segmentation output is a coherent region with connected edges. Simplicity in terms of concept and procedures is an advantage of region growing for immediate implementation.

Region-based segmentation is insensitive to image semantics; it does not recognize object but only predefined membership function. Besides, the design of the region membership is as difficult as setting a threshold value; region-based segmentation is unable to separate multiple disconnected objects simultaneously. The assumption that the region within a group of object is homogenous has low practical value in hand bone segmentation due to the fact that the bone is formed by cancellous bone and cortical bone that has high variations on texture and intensity range. Besides, in the presence of noise or any unexpected variations, region growing leads to holes or extra-segmented region in the resultant segmented region and thus has low accuracy in certain condition (Mehnert and Jackway 1997). The number and the location of seeds and membership function in seeded region growing, as well as the merging criteria in split–merge region growing, depend on human decisions which are subjective and laborious.

One of the famous region growing methods is the split and merge algorithm; split and merge is an algorithm splitting the image successively until a specified number of regions remain (Tremeau and Borel 1997). To perform the split and merge region growing algorithm, firstly, the entire image is considered within one region. Then, the splitting process begins in the region in accordance with the homogeneity criterion; if the criterion is met, then it splits (Gonzalez and Woods 2007). This splitting process repeats until all regions are homogenous. After the splitting process, the merging process begins. Initially, comparison among neighborhood regions is performed. Then, the region merges to each other according to some criterion such as the pixels' intensity value where regions that are less than the standard deviation are considered homogenous.

We have reviewed the essential concept of region-based segmentation. The purpose is to identify coherent regions defined by pixel similarities. The main challenge of this type of segmentation is often related to the pixel similarities: what are the features that should be adopted as similarities measurement and how are the thresholds of chosen features should be set in defining the similarity. The selection of features is difficult as they depend on application. For example, if the targeted object is not a connected object, pixel intensity is not suitable as pixel similarities measurement. The setting of threshold is another tricky challenge as it manipulates the trade-offs in terms of flexibility. For example, if the threshold is set too low, the inferior effect of over-segmentation occurs because pixels easily surpass the threshold leading to larger coherent regions than the actual objects; if the threshold is set too high, the otherwise occurs. Region-based segmentation is unable to segment objects of multiple disconnected regions, and therefore, in the context of hand bone segmentation, applying only region-based segmentation is inappropriate as children hand bones for BAA involve different numbers of bones regions at different ages.

Deformable model refers to classes of methods that implement an estimated model of the targeted object using the model constructed by the prior information such as the texture and shape variability of the specific class of object as flexible two-dimensional curves or three-dimensional surfaces. In two-dimensional cases, these curves deform elastically to by satisfying some constraints to match the borders of the targeted object in a given image. The word 'active' stems primarily from the nature of the curves in adapting themselves to fit the targeted object. There are three main classes of deformable model: active contour model, active shape model, and AAM.

Deformable models assemble the mathematical knowledge from physics in limiting the shape flexibility over the space, geometry in shape representation, and optimization theory in model-object fitting. These mathematical foundations work together by playing their roles to establish the deformable model. For instance, the geometric representation with certain degree of freedoms is to cover broader shape changes; the principle in physics, in accordance with forces and constraints, controls the changes of shape to permit only meaningful geometric flexibility; optimization theory adjusts the shape to fulfill the objective function constituted by external energy and internal energy; the external energy is associated with the deformation of model to fit the targeted object due to external potential energy, whereas the internal energy constrains the smoothness of the constructed model in terms of internal elasticity forces.

Kass et al. (1988) proposed Active contour model or known as 'snake' as a potential solution to segmentation problem (Leymarie 1986). From the perspective of geometry, it is an embedded parametric curve represented as $v(s) = (x(s), y(s))^T$ on image plane $(x, y) \in R^2$, where $x(.)$ and $y(.)$ denote coordinates functions, and $s \in [0, 1]$ denotes the parametric domain. A snake in this context illustrates an elastic contour that fits to some preferred features in image.

To apply active contour model in segmentation, first, establish the initial location of point s in image planes adjacent to targeted object. These points collect 'evidence' locally in their territories and feedback to the contour energy. Next, search the update of each point using local information by solving the Euler–Lagrange equation when the contour is in equilibrium according to calculus of variation. Conventionally, numerical algorithm is applied to solve the equation in discrete approximation framework. Lastly, these steps repeat until stopping criteria has been achieved.

Since the active contour model is proposed, a lot of variations have been introduced by scholars. We have summarized some of them which are highly cited as following:

The advantages of active contour compared with previously discussed methods:

1. Process the image pixels in specific areas only instead of the entire image and thus enhance the computational efficiency.
2. Impose certain controllable prior information.
3. Impose desired properties, for instance, contour continuity and smoothness.
4. Can be easily governed by user by manipulating the external forces and constraints.
5. Respond to image scale accordingly with the assistance of filtering process.

Disadvantages of classical active contour model

1. Not specific enough to be implemented in specific problem as the shape of the targeted object is often not recognized by the algorithm.
2. Unable to segment multiple objects.
3. High sensitivity to environmental noises in image.
4. High dependency toward intensity gradient along the edges.
5. Do not consider the region information of the targeted objects.
6. High dependency on initial guessed point location. If the initial snake is not sufficiently close to targeted object boundaries, then points in snake can hardly attach the boundaries.
7. Difficult to grow into concavity.
8. Do not have a global shape controller that constraints the shape of contour from deviating from allowable shape of the targeted object.

2.5.1 Balloon Snake

Classical snake suffers from two drawbacks. The first drawback is that the contour model shrinks by searching a point of equilibrium based on the internal energy and boundary conditions if external energy is absent; the second drawback is that a contour model that is not close enough to the targeted object boundaries, the attraction of the model to the boundaries is very low. Balloon snake attempt to solve the problem by introducing the inflating force that makes the contour model behaves similarly as a balloon in two dimensions so that the contour model stops not at spurious edges by considering edges point extracted from Canny-Deriche edge detector (Cohen 1991; Cohen and Cohen 1993).

2.5.2 Level Set

It is first introduced by Osher and Sethian (1988) in the area of fluid dynamics, then being applied in computer vision for segmentation by Caselles et al. (1993) and Malladi et al. (1995). Later, this method has been incorporated with region information and boundary information by Paragios and Deriche (1999). The level set methods, different from classical snake, model the contour in terms of implicit surface extracted from initial curve and then establish the connection between the curve propagation flow and implicit function deformation flow. The curvature and image gradient are then used to evolve the surface.

2.5.3 Active Contour Without Edges

Chan-Vese snake (Chan and Vese 2001) is an exceptionally popular extension of level set by combining level set method with Mumford-Shah functional segmentation technique (Mumford and Shah 1989). The prime utility of this snake is that it can detect objects even without using its gradient information. They minimize the energy in the level set formulation to evolve a curve attaching to boundary in such a way that the stopping term relies not on the boundary, the final result requires no smoothing procedure, and the initial contour need not to be around the targeted object.

2.5.4 Geodesic Snake

Classical snake often faces problems associated with incapability to detect multiple objects and incapability to detect interior and exterior boundaries simultaneously. An extension to level set snake, Geodesic snake (Caselles et al. 1997), with the implementation of geodesic computational approach, curve evolution theory, and geometric flows, improve the contour models so that they can split and merged without additional prior information or additional topology processes. This geodesic snake is then being upgraded by Leventon et al. (2002) by incorporating prior shape model. Firstly, analysis of variance of a set of shapes is performed. Then, maximum a posterior (MAP) position is estimated at every evolution step in curve. This extension has improved the robustness of boundary convergence in active contour despite noisy inputs.

2.5.5 Gradient Vector Flow Snake

The limited respond of classical snake in contour initiation and convergence associated with concavity has been addressed by GVF snake by applying an external field extracted from diffusion of gradient vectors derived from gray level of binary image (Xu and Prince 1997, 1998a, b) The main idea is to diffuse the forces to far situated contour model from the object by minimizing the an energy functional after solving two coupled partial differential equations. This is important to attract contour model that is initiated far from targeted object and can at the same time resolve the problem of discontinuity of object boundaries. Besides, the GVF produces forces in large capture range to hence able to attract and progress contour into concavity.

To sum up, the regularizing terms adopted in active contour model is useful in stabilizing the contour, but the robustness is limited as the imposed constraints generally tend to smooth and shorten the contour unless stronger external energy is involved; this scheme is often too general and inadequate. Therefore, a more specifically designed scheme that capable of incorporating more finely tuned prior knowledge about the class of targeted object is required and is explained in next subsection.

Active shape model is a model founded on statistical theory where the variations of the shape of the objects can be captured via training procedure using labeled object's contour in the image in set points representation. Activating the trained contour will deform the contour fitting the targeted object in the image. Cootes et al. (1995) developed the model. Generally, it works by searching the best position of initial points that are surrounding the object and then updating these positions until the stopping criteria are achieved through iterations. Ever since the technique is proposed until recently, it has been extensively applied in various fields such as facial recognition (Xue et al. 2003; Zheng et al. 2008; Sukno et al. 2010), object tracking (Jang and Choi 2000; Kim and Lee 2005; Nuevo et al. 2011; Liu et al. 2012), and medical image processing (Smyth et al. 1997; Hodge et al. 2006; Aung et al. 2010; Toth et al. 2011).

The deformable models, ASM and AAM, undoubtedly are powerful segmentation methods. However, they are not without weaknesses and are not best method for automated hand segmentation. The reasons are summarized as following:

1. The landmarks placement has to be manually annotated by users. Incorrect landmarks placements lead to unreliable capture of shape variability.
2. The number of landmarks has to be specified by user manually. Insufficient landmarks lead to failure in obtaining the shape of the targeted structures; excessive landmarks lead to computational inefficiency.
3. The training phase requires a lot of training examples in database which is not necessarily available in many applications. Insufficient training examples lead to failure in generalizing the mean structure's shape.
4. The nature of hand bone development of children: different numbers and sizes of bones in different ranges of age complicated the implementation ASM and AAM especially in establishing the general form of mean shape.
5. The alignment phase is uncertain in terms of its numerical stability: the convergence of the mean model in the iterative method has not been devised mathematically and prone to errors.
6. The choice in retaining the number of eigenvectors in principal component analysis has to be determined correctly by user. Incorrect decision leads to failure in capturing the representative points of the shape; consequently, inaccurate model is constructed and leads to undesired segmentation result.

7. Variations in hand structural positions are often largely deviated and this devotes to nonlinear parameter relations that invariably impede the accurate segmentation as a whole.

ASM has been applied by Thodberg and Rosholm (2003) to address the problem of hand bone segmentation. Extensive training has to be done to complete the model in order to imitate the recognition understanding of human beings in segmenting the hand bone. Note that the initiation of set points placement to mark the spatial position of hand bone shape demands expert to be the operators. Both requirements of training set and human expert are the main weakness of this model in addressing the problem. It is tedious, subjective and time-consuming to delineate the shape from a large training set, not to mention the critical issue of the availability of these resources. Therefore, an alternative segmentation framework has to be established when the resources are limited and this motivates the research of this thesis.

AAM is a statistical model of shape and gray-level appearance of the targeted object proposed by Edwards et al. (1998). The final aim is to generalize the model to all valid example (Cootes et al. 1996). The relationship between the model parameter displacements and the errors between training example and a model instance is learned during the training phase. By computing the errors of fitting and using the previously obtained parameters, the current parameters with the intent of improving the current fitting can be updated.

AAMs and the closely related concepts are found in the methods of active shape model. The AAMs are most frequently being adopted in the application related to face modeling. Besides face modeling (Butakoff and Frangi 2010), it has been implemented in other applications as well such as in medical image processing (Roberts et al. 2007; Patenaude et al. 2011). The typical first step of AAM is to fit the AAM to an input image using model parameters that maximize the matching criteria between the model instance and the input image. The model parameters are then passed to a classifier to yield t classification tasks.

To fit the AAM to an input image involves solving an nonlinear optimization problem. The conventional method of solving the problem is by updating the parameters iteratively. This update has to be incremental additive and the parameters refer to shape and appearance coefficients. The input image can be warped onto the model coordinate frame by using the current shape parameters estimations. The error between the model instance and the fitting of AAM onto the image can be computed. This error is then acted as feedback in next iteration that would affect the updates of the parameters. The constant coefficients in this linear relationship between the updates and errors can then be found either by linear regression or by other numerical methods.

Although the AAM appears to be the useful model-based system in medical image segmentation, it has constraints that impede its performance in practical application (Gao et al. 2010).

1. Low efficiency in real-time systems: current algorithm of AAM consumes a lot of time and space computational costs. Thus, it is of prime importance to minimize the complexities in time and space needed to perform the algorithm in order to realize it in real-time system. The efficiency is mainly affected by the following factors: manual landmarks placement, complex texture representation in high-resolution medical image, iterative procedure in solving the optimization problem.

2. Low discriminative ability for recognition and segmentation systems: only a group of object is being modeled, and thus it is considered as a generative model which possesses no ability to classify different objects. This ability depends on the accuracy of model fitting which are affected by how the prior shape is chosen; how the texture is represented; how the texture is modeled. It is crucial to improve this discriminative ability to perform segmentation tasks effectively.

3. Inconsistent robustness under different circumstances: the performance of the system is influenced by different conditions such as the existence of pose variations, uneven illumination, the absence of features, low resolution, and the presence of noises.

AAM is a very useful model as it can capture the mode of variations of deformable objects given a set of training examples. The mode of variations includes shape and texture as a whole. Besides, it can perform the projection of object onto low-dimensional subspace to reduce redundancy and capture main component of variations. Thus, it has been implemented in a lot of applications especially medical image segmentation. Nonetheless, it has limitation in efficiency, discriminative ability, and robustness in different condition. In the problem of hand bone segmentation, the same group of researchers that adopted ASM has extended their works by applying AAM (Thodberg 2002; Thodberg et al. 2010). The weaknesses discussed in applying ASM remain because the technical differences between AAM and ASM enhance only the robustness in terms of prior knowledge and the information around the object that have been incorporated into the model, not the practicality in terms of availability of training set and expert operators.

2.6 Desired Properties of Segmentation

Top-down strategy is adopted in designing the proposed segmentation framework. Firstly, the overview of the desired system is obtained through literature reviews by reviewing the existing techniques and analyzing the factors leading them to failure as effective hand bone segmentation technique. After gaining some insights into constituting a desired hand bone segmentation framework, we then identify the desired characteristics, only then we propose each sub-framework to satisfy each requirement.

The desired characteristics of hand bone segmentation framework adopted in computerized BAA should comprise the followings:

1. Contrast, illumination, orientation invariance: to ensure consistent segmentation robustness under different conditions of X-ray settings and devices.
2. Relatively low computational complexity: to ensure practical execution time for automated BAA system. Ideally, it is comprehensive enough to tackle with image complexities and uncertainties, yet it is simple enough to be executed in a reasonable time frame.
3. No complicated 'training' procedures: to ensure no dependency on availability of training samples of hand bone radiographs. However, simple parameter tuning procedures without depending on availability of training hand bone radiographs have to be established to capture the variations of uncertainties in image nature.
4. Utilization of prior knowledge: to ensure the usage of available information to optimize the result on hand bone segmentation. Besides, making use of 'by-products' of image preprocessing is preferable.
5. Relatively high resistance to noises: to ensure good performance of segmentation despite the inevitable random signals in the hand bone radiographs.
6. Automated or minimum dependency on human interventions: to ensure objectivity, to enable reproducibility, and to avoid laboriousness.
7. Consistent accuracy: to ensure relatively high precision in segmentation on resultant hand bone for subsequent processing in automated BAA system.
8. Resemblance to manual segmentation: to ensure a certain level of artificial intelligence in the designed algorithm to emulate human visual perception.
9. No overdependence on certain image feature: to ensure segmentation robustness under the absence of any certain property such as intensity discontinuity or edges.
10. Adaptability: to ensure robustness under the presence of variability in different regions of hand radiographs.
11. Optimality: all parameters are chosen based on the direction of finding the optimum solution and not arbitrarily preset. However, this criterion should not violate the second criterion.

To facilitate the subsequent explanations on our propose framework, henceforth, aforementioned desired property is referred as P1, P2, P3 … and so forth. For example, the first property of contrast, illumination, orientation invariance is referred as P1 and the tenth criterion of adaptability is referred as P10.

References

Acheson RM, Fowler G, Fry EI, Janes M, Koski K, Urbano P, Werfftenboschjj VA (1963) Studies in the reliability of assessing skeletal maturity from x-rays: part III. Greulich-Pyle Atlas and Tanner-Whitehouse method contrasted. Hum Biol Int Rec Res 35:317–349
Aicardi G, Vignolo M, Milani S, Naselli A, Magliano P, Garzia P (2000) Assessment of skeletal maturity of the hand-wrist and knee: a comparison among methods. Am J Hum Biol 12(5):610–615

Aja-Fernández S, De Luis-García R, Martín-Fernández MÁ, Alberola-López C (2004) A computational TW3 classifier for skeletal maturity assessment. A Computing with Words approach. J Biomed Inform 37(2):99–107

Aung MSH, Goulermas JY, Stanschus S, Hamdy S, Power M (2010) Automated anatomical demarcation using an active shape model for videofluoroscopic analysis in swallowing. Med Eng Phys 32(10):1170–1179

Baradez MO, McGuckin CP, Forraz N, Pettengell R, Hoppe A (2004) Robust and automated unimodal histogram thresholding and potential applications. Pattern Recogn 37(6):1131–1148

Bernsen J (1986) Dynamic thresholding of grey-level images. IEEE, pp 1251–1255

Buie HR, Campbell GM, Klinck RJ, MacNeil JA, Boyd SK (2007) Automatic segmentation of cortical and trabecular compartments based on a dual threshold technique for in vivo micro-CT bone analysis. Bone 41(4):505–515

Butakoff C, Frangi AF (2010) Multi-view face segmentation using fusion of statistical shape and appearance models. Comput Vis Image Underst 114(3):311–321

Cao F, Huang HK, Pietka E, Gilsanz V (2000) Digital hand atlas and web-based bone age assessment: system design and implementation. Comput Med Imaging Graph 24(5):297–307

Caselles V, Catté F, Coll T, Dibos F (1993) A geometric model for active contours in image processing. Numer Math 66(1):1–31

Caselles V, Kimmel R, Sapiro G (1997) Geodesic active contours. Int J Comput Vision 22(1):61–79

Chan TF, Vese LA (2001) Active contours without edges. IEEE Trans Image Process 10(2):266–277

Chemaitilly W, Sklar CA (2010) Endocrine complications in long-term survivors of childhood cancers. Endocr Relat Cancer 17(3):R141–R159

Cohen LD (1991) On active contour models and balloons. CVGIP: Image Underst 53(2):211–218

Cohen LD, Cohen I (1993) Finite-element methods for active contour models and balloons for 2-D and 3-D images. IEEE Trans Pattern Anal Mach Intell 15(11):1131–1147

Cootes TF, Page GJ, Jackson CB, Taylor CJ (1996) Statistical grey-level models for object location and identification. Image Vis Comput 14(8):533–540

Cootes TF, Taylor CJ, Cooper DH, Graham J (1995) Active shape models-their training and application. Comput Vis Image Underst 61(1):38–59

De Santis A, Sinisgalli C (1999) A Bayesian approach to edge detection in noisy images. Circ Syst I: Fundamental Theory Appl, IEEE Trans 46(6):686–699

Edwards GJ, Taylor CJ, Cootes TF (1998) Interpreting face images using active appearance models. In: FG'98. Proceedings of the 3rd international conference on face and gesture recognition, IEEE computer society

Gao X, Su Y, Li X, Tao D (2010) A review of active appearance models. Syst, Man, Cybern, Part C: Appl Rev, IEEE Trans 40(2):145–158

Giordano D, Spampinato C, Scarciofalo G, Leonardi R (2010) An automatic system for skeletal bone age measurement by robust processing of carpal and epiphysial/metaphysial bones. IEEE Trans Instrum Meas 59(10):2539–2553

Gonzalez R, Woods R (2007) Digital Image Processing, 3rd edn. Prentice Hall, Upper Saddle River

Greulich W, Pyle S (1959) Radiographic atlas of skeletal development of hand wrist. Am J Med Sci 238(3):393

Guo R, Pandit SM (1998) Automatic threshold selection based on histogram modes and a discriminant criterion. Mach Vision Appl 10(5–6):331–338

Han C-C, Lee C-H, Peng W-L (2007) Hand radiograph image segmentation using a coarse-to-fine strategy. Pattern Recogn 40(11):2994–3004

Heinrich UE (1986) Significance of radiologic skeletal age determination in clinical practice. Die Bedeutung der radiologischen Skelettalterbestimmung für die Klinik 26(5):212–215

Hodge AC, Fenster A, Downey DB, Ladak HM (2006) Prostate boundary segmentation from ultrasound images using 2D active shape models: optimisation and extension to 3D. Comput Methods Progr Biomed 84(2–3):99–113

Hsieh CW, Jong TL, Chou YH, Tiu CM (2007) Computerized geometric features of carpal bone for bone age estimation. Chin Med J 120(9):767–770

Hsieh CW, Jong TL, Tiu CM (2008) Carpal growth assessment based on fuzzy description. In: Soft computing in industrial applications, 2008. SMCia'08. IEEE conference on. pp 355–358

Huang Q, Gao W, Cai W (2005) Thresholding technique with adaptive window selection for uneven lighting image. Pattern Recogn Lett 26(6):801–808

Hum YC (2013) Segmentation of hand bone for bone age assessment. Springer, London. Limited, 2013. ISBN: 9814451657, 9789814451659. 125 pp (SpringerBriefs in Applied Sciences and Technology Series)

Jang D-S, Choi H-I (2000) Active models for tracking moving objects. Pattern Recogn 33(7):1135–1146

Jong-Min L, Whoi-Yul K (2008) Epiphyses extraction method using shape information for left hand radiography. In: Convergence and hybrid information technology, 2008. ICHIT '08. International conference on, 28–30 Aug 2008, pp 319–326

Jonsson K (2002) Fundamentals of hand and wrist imaging. Acta Radiol 43(2):236

Kang C-C, Wang W-J, Kang C-H (2012) Image segmentation with complicated background by using seeded region growing. AEU Int J Electron Commun 66(9):767–771

Kass M, Witkin A, Terzopoulos D (1988) Snakes: active contour models. Int J Comput Vision 1(4):321–331

Kim W, Lee J–J (2005) Object tracking based on the modular active shape model. Mechatronics 15(3):371–402

Lee SU, Yoon Chung S, Park RH (1990) A comparative performance study of several global thresholding techniques for segmentation. Comput Vision, Graph, Image Process 52(2):171–190

Leventon ME, Grimson WEL, Faugeras O (2002) Statistical shape influence in geodesic active contours. Computer vision and pattern recognition, 2000. In: Proceedings. IEEE conference on. pp 316–323

Leymarie FF (1986) Tracking and desribing deformable objects using active contour models. Master thesis, McGill University

Liu J, Qi J, Liu Z, Ning Q, Luo X (2008) Automatic bone age assessment based on intelligent algorithms and comparison with TW3 method. Comput Med Imaging Graph 32(8):678–684

Liu Z, Shen H, Feng G, Hu D (2012) Tracking objects using shape context matching. Neurocomputing 83:47–55

Liyuan L, Ran G, Weinan C (1997) Gray level image thresholding based on fisher linear projection of two-dimensional histogram. Pattern Recogn 30(5):743–749

Luijendijk H (1991) Automatic threshold selection using histograms based on the count of 4-connected regions. Pattern Recogn Lett 12(4):219–228

Mahmoodi S, Sharif BS, Chester EG, Owen JP, Lee REJ (1997) Automated vision system for skeletal age assessment using knowledge based techniques. IEE, pp 809–813

Mahmoodi S, Sharif BS, Chester EG, Owen JP, Lee REJ (1999) Bayesian estimation of growth age using shape and texture descriptors. IEE, pp 489–493

Mahmoodi S, Sharif BS, Graeme Chester E, Owen JP, Lee R (2000) Skeletal growth estimation using radiographie image processing and analysis. IEEE Trans Inf Technol Biomed 4(4):292–297

Malladi R, Sethian JA, Vemuri BC (1995) Shape modeling with front propagation: a level set approach. Pattern Anal Mach Intell, IEEE Trans 17(2):158–175

Manos G, Cairns AY, Ricketts IW, Sinclair D (1993) Automatic segmentation of hand-wrist radiographs. Image Vis Comput 11(2):100–111

Manos GK, Cairns AY, Rickets IW, Sinclair D (1994) Segmenting radiographs of the hand and wrist. Comput Methods Progr Biomed 43(3–4):227–237

Martin DD, Deusch D, Schweizer R, Binder G, Thodberg HH, Ranke MB (2009) Clinical application of automated Greulich-Pyle bone age determination in children with short stature. Pediatr Radiol 39(6):598–607

Martin DD, Heckmann C, Jenni OG, Ranke MB, Binder G, Thodberg HH (2011) Metacarpal thickness, width, length and medullary diameter in children-reference curves from the First Zürich Longitudinal Study. Osteoporos Int 22(5):1525–1536

Mehnert A, Jackway P (1997) An improved seeded region growing algorithm. Pattern Recogn Lett 18(10):1065–1071

Michael DJ, Nelson AC (1989) HANDX: a model-based system for automatic segmentation of bones from digital hand radiographs. IEEE Trans Med Imaging 8(1):64–69

Mumford D, Shah J (1989) Optimal approximations by piecewise smooth functions and associated variational problems. Commun Pure Appl Math 42(5):577–685

Niblack W (1990) An introduction to digital image processing. Prentice Hall, Upper Saddle River

Niemeijer M, Van Ginneken B, Maas CA, Beek FJA, Viergever MA (2003) Assessing the skeletal age from a hand radiograph: automating the tanner-whitehouse method. In: Proceedings of the 2003 SPIE medical imaging. pp 1197–1205

Nuevo J, Bergasa LM, Llorca DF, Ocaña M (2011) Face tracking with automatic model construction. Image Vis Comput 29(4):209–218

Ontell FK, Ivanovic M, Ablin DS, Barlow TW (1996) Bone age in children of diverse ethnicity. Am J Roentgenol 167(6):1395–1398

Osher S, Sethian JA (1988) Fronts propagating with curvature-dependent speed: algorithms based on Hamilton-Jacobi formulations. J Comput Phys 79(1):12–49

Paragios N, Deriche R (1999) Geodesic active regions for supervised texture segmentation. In: Proceedings of the international conference on computer vision, vol 2, IEEE computer society. p 926

Patenaude B, Smith SM, Kennedy DN, Jenkinson M (2011) A Bayesian model of shape and appearance for subcortical brain segmentation. NeuroImage 56(3):907–922

Peloschek P, Nemec S, Widhalm P, Donner R, Birngruber E, Thodberg HH, Kainberger F, Langs G (2009) Computational radiology in skeletal radiography. Eur J Radiol 72(2):252–257

Pietka E (1994) Computer-assisted bone age assessment based on features automatically extracted from a hand radiograph. Comput Med Imaging Graph 19(3):251–259

Pietka E, Kaabi L, Kuo ML, Huang HK (1993) Feature extraction in carpal-bone analysis. IEEE Trans Med Imaging 12(1):44–49

Pietka E, Gertych A, Pospiech S, Cao F, Huang HK, Gilsanz V (2001) Computer-assisted bone age assessment: Image preprocessing and epiphyseal/metaphyseal ROI extraction. IEEE Trans Med Imaging 20(8):715–729

Roberts M, Cootes T, Pacheco E, Adams J (2007) Quantitative vertebral fracture detection on DXA images using shape and appearance models. Acad Radiol 14(10):1166–1178

Roche AF, French NY (1970) Differences in skeletal maturity levels between the knee and hand. Am J Roentgenol 109(2):307–312

Sebastian TB, Tek H, Crisco JJ, Kimia BB (2003) Segmentation of carpal bones from CT images using skeletally coupled deformable models. Med Image Anal 7(1):21–45

Sezgin M, Sankur B 1 (2004) Survey over image thresholding techniques and quantitative performance evaluation. J Electron Imaging 13(1):146–168

Shafait F, Keysers D, Breuel TM (2008) Efficient implementation of local adaptive thresholding techniques using integral images. Proceedings of SPIE 6815, Document Recognition and Retrieval XC, 681510 (January 28, 2008)

Shapiro L, Stockman G (2001) Computer vision. Prentice Hall, Upper Saddle River

Sharif BS, Zaroug SA, Chester EG, Owen JP, Lee EJ (1994) Bone edge detection in hand radiographic images. In: IEEE. pp 514–515

Smyth PP, Taylor CJ, Adams JE (1997) Automatic measurement of vertebral shape using active shape models. Image Vis Comput 15(8):575–581

Somkantha K, Theera-Umpon N, Auephanwiriyakul S (2011a) Bone age assessment in young children using automatic carpal bone feature extraction and support vector regression. J Digit Imaging 24(6):1044–1058

Somkantha K, Theera-Umpon N, Auephanwiriyakul S (2011b) Boundary detection in medical images using edge following algorithm based on intensity gradient and texture gradient features. IEEE Trans Biomed Eng 58(3 PART 1):567–573

Sotoca JM, Iñesta JM, Belmonte MA (2003) Hand bone segmentation in radioabsorptiometry images for computerised bone mass assessment. Comput Med Imaging Graph 27(6):459–467

Sukno FM, Guerrero JJ, Frangi AF (2010) Projective active shape models for pose-variant image analysis of quasi-planar objects: application to facial analysis. Pattern Recogn 43(3):835–849

Tanner J, Whitehouse R (1975) Assessment of skeletal maturity and prediction of adult height (TW2 method)

Tanner JM (2001) Assessment of skeletal maturity and prediction of adult height (TW3 method). W.B. Saunders, London

Tanner JM, Gibbons RD (1994) Automatic bone age measurement using computerized image analysis. J Pediatr Endocrinol 7(2):141–145

Thodberg HH (2002) Hands-on experience with active appearance models. pp 495–506

Thodberg HH, Jenni OG, Caflisch J, Ranke MB, Martin DD (2009a) Prediction of adult height based on automated determination of bone age. J Clin Endocrinol Metab 94(12):4868–4874

Thodberg HH, Kreiborg S, Juul A, Pedersen KD (2009b) The BoneXpert method for automated determination of skeletal maturity. Med Imaging, IEEE Trans 28(1):52–66

Thodberg HH, Rosholm A (2003) Application of the active shape model in a commercial medical device for bone densitometry. Image Vis Comput 21(13–14):1155–1161

Thodberg HH, Van Rijn RR, Tanaka T, Martin DD, Kreiborg S (2010) A paediatric bone index derived by automated radiogrammetry. Osteoporos Int 21(8):1391–1400

Toth R, Tiwari P, Rosen M, Reed G, Kurhanewicz J, Kalyanpur A, Pungavkar S, Madabhushi A (2011) A magnetic resonance spectroscopy driven initialization scheme for active shape model based prostate segmentation. Med Image Anal 15(2):214–225

Tran Thi My Hue MGS, Kim JY, Choi SH (2011) Hand bone image segmentation using watershed transform with multistage merging. J Korean Instit Inf Technol 9(5): 59–66

Tremeau A, Borel N (1997) A region growing and merging algorithm to color segmentation. Pattern Recogn 30(7):1191–1203

Tristán-Vega A, Arribas JI (2008) A radius and ulna TW3 bone age assessment system. IEEE Trans Biomed Eng 55(5):1463–1476

Tsai D-M (1995) A fast thresholding selection procedure for multimodal and unimodal histograms. Pattern Recogn Lett 16(6):653–666

Whatmough RJ (1991) Automatic threshold selection from a histogram using the "exponential hull". CVGIP: Graph Models Image Process 53(6):592–600

Xu C, Prince JL (1997) Gradient vector flow: a new external force for snakes. IEEE, pp 66–71

Xu C, Prince JL (1998a) Generalized gradient vector flow external forces for active contours. Sig Process 71(2):131–139

Xu C, Prince JL (1998b) Snakes, shapes, and gradient vector flow. IEEE Trans Image Process 7(3):359–369

Xue Z, Li SZ, Teoh EK (2003) Bayesian shape model for facial feature extraction and recognition. Pattern Recogn 36(12):2819–2833

Yan F, Zhang H, Kube CR (2005) A multistage adaptive thresholding method. Pattern Recogn Lett 26(8):1183–1191

Zhang A, Gertych A, Liu BJ (2007) Automatic bone age assessment for young children from newborn to 7-year-old using carpal bones. Comput Med Imaging Graph 31(4–5):299–310

Zhang J, Huang HK (1997) Automatic background recognition and removal (ABRR) in computed radiography images. IEEE Trans Med Imaging 16(6):762–771

Zhao M, Yang Y, Yan H (2000) An adaptive thresholding method for binarization of blueprint images. Pattern Recogn Lett 21(10):927–943

Zheng Z, Jiong J, Chunjiang D, Liu X, Yang J (2008) Facial feature localization based on an improved active shape model. Inf Sci 178(9):2215–2223

Chapter 3
Review on Advanced Techniques in 2-D Fetal Echocardiography: An Image Processing Perspective

Dyah Ekashanti Octorina Dewi, Heamn Noori Abduljabbar and Eko Supriyanto

Abstract Vast advancement of digital ultrasound imaging technology in diagnostic and therapeutic purposes has benefitted from the large development of image processing systems. The conjugation of these two fields has supported the ultrasound imaging system with sophisticated data analysis and visualization in a more lucid and interactive way and opened wide opportunities to new ultrasound imaging applications. In obstetric care, two-dimensional (2-D) fetal echocardiography has become main routine procedure to evaluate cardiac status of the fetus so that any cardiac anomalies can be detected earlier. Seeing that fetal cardiology has its functional uniqueness, fetal echocardiography has its specific assessment techniques that differ from other purposes. Furthermore, the existing conventional fetal echocardiography techniques are found to have some limitations that hamper the accuracy of the assessment. On the other hand, the current rapid development of computer technology in medical imaging, especially in fetal echocardiography, has been proven to improve the image quality and measurement accuracy of the fetal heart imaging. This chapter provides a review of 2-D fetal echocardiography techniques in image processing point of view. We concentrate our review on three main aspects to support diagnosis in 2-D fetal echocardiography: speckle reduction, image segmentation, and image analysis. This review is mainly aimed at appraising several advanced techniques in fetal echocardiography and finding possibilities for future developments in both clinical and research fields.

D. E. O. Dewi · H. N. Abduljabbar · E. Supriyanto (✉)
Faculty of Biosciences and Medical Engineering, IJN-UTM Cardiovascular Engineering Center, Universiti Teknologi Malaysia, Skudai-Johor, Malaysia
e-mail: eko@biomedical.utm.my

D. E. O. Dewi
e-mail: dyah.ekashanti@gmail.com

H. N. Abduljabbar
e-mail: heamn_jabbari@yahoo.com

K. W. Lai et al., *Advances in Medical Diagnostic Technology*,
Lecture Notes in Bioengineering, DOI: 10.1007/978-981-4585-72-9_3,
© Springer Science+Business Media Singapore 2014

Keywords Two-dimensional fetal echocardiography · Congenital heart disease · Cardiac magnetic resonance · Image processing · Speckle reduction · Image segmentation · Image analysis

3.1 Introduction

Congenital anomalies (also referred as birth defects, congenital disorders, or congenital malformations) are mostly related to cardiac defects, neural tube defects, and Down syndrome (World Health Organization 2012; Lowry et al. 2013). The congenital heart disease (CHD) problem, as the most common severe congenital anomaly found in neonates, has significant impact not only on infant morbidity and mortality, but also on health care cost in children and adults. Although the etiology is exactly unknown, the incidence of CHD is closely related to other associated major and minor non-cardiac anomalies. This is mostly associated with potential genes and molecular mechanisms, exogenous, and other multifactorial factors that affect embryonic development phase, such as maternal risk and environmental factors (Gembruch 1997; Sander et al. 2006; Baardman et al. 2012). Therefore, CHD cases necessitate to be managed by prenatal screening procedure to improve the survival rate of the fetus. This is realized by performing an in-depth fetal cardiovascular evaluation by using fetal imaging device (Gardiner 2001; Garne et al. 2001).

In the literature, to the authors' knowledge, fetal heart can only be monitored by using cardiac magnetic resonance (CMR) and fetal echocardiography (Sklansky 2004; Prakash et al. 2010). Fetal echocardiography, as the main obstetric imaging screening for fetus, is considered to be the most effective fetal monitoring system with excellent diagnostic accuracy to diagnose cardiac anomalies prenatally (Meyer-Wittkopf et al. 2001; Chew et al. 2007; Deng and Rodeck 2004), due to its non-radiation exposure, real-time capability, economic rate, and ease of access. The main purpose of fetal echocardiography is to perform assessment of fetal cardiac structure and function, detect complex structural malformation in early stage, and recognize the presence of a defect by using a specialized sonography system. With the combination of expertise knowledge and imaging instrument, most forms of major heart disease have become recognizable by prenatal ultrasound (World Health Organization 2012; Allan et al. 1994; Jaeggi et al. 2001; Lagopoulos et al. 2010). Such screening examination is proven to optimize fetal cardiac monitoring by identifying and characterizing the progression of congenital anomalies before delivery and to aid management decisions in delivery, treatment, or interventions (Garne et al. 2001; Levi 2002; Carvalho 2002; Lee 2013; Carvalho et al. 2013; Ayres 1997; Chitty and Pandya 1997).

In this review, we attempt to provide supplementary information about technical aspects in fetal echocardiography by bringing up the discussion into newer issues in image processing and analysis methods for developing 2-D fetal echocardiography applications. We also restrict our discussion specifically in 2-D fetal echocardiography for diagnostic purpose only. Therefore, topics related to interventional echocardiography is not included in this review. We organize the article by presenting a brief introduction about general aspects in fetal echocardiography. A comparison between CMR and fetal echocardiography is also given here. Section 3.2 reviews about fetal heart imaging modalities, CMR and fetal echocardiography, including the new developments. In Sect. 3.3, a brief overview on basic clinical routine and standards in 2-D fetal echocardiography is portrayed. Furthermore, a thorough discussion on imaging and image processing to solve problems in 2-D fetal echocardiography images is revealed. The discussion includes transducers and settings, speckle reduction, image segmentation, and image analysis. Furthermore, a concise evaluation of these techniques is given to describe the advantages and drawbacks of the techniques. Finally, possible future directions for clinical and research applications are proposed.

3.2 Fetal Heart Imaging

The major issue in choosing fetal heart imaging is about noninvasive modality. Although echocardiography remains the core of noninvasive cardiac imaging, the existence of CMR is potential to take part in improving the quality of the diagnosis (Prakash et al. 2010). In this section, we perform a simple comparison study of the two modalities, CMR and echocardiography, to characterize the diagnosis constraints.

3.2.1 Cardiac Magnetic Resonance (CMR): A Brief Overview

CMR, as the branch of magnetic resonance imaging (MRI) that specializes in cardiovascular imaging, also has the potency to perform fetal heart imaging. Hitherto, CMR has been greatly implemented to scan CHD in infants, children, and adults, especially in the anatomic assessment of cardiovascular anomalies before and after surgery, quantification of biventricular function, magnetic resonance angiography (MRA), measurement of systemic and pulmonary blood flow, quantification of valve regurgitation, identification of myocardial ischemia and fibrosis, and tissue characterization (Prakash et al. 2010; Pennell 2010; Tsai-Goodman et al. 2004; Chung 2000; Khattab et al. 2013; Orwat et al. 2013). Although CMR is considered to be new concept in fetal cardiac imaging, several studies have shown its bright prospect for further development. Initially, fetal CMR was only applied in postmortem and in vitro studies (Deng and Rodeck 2004; Fenton et al. 2001; Kurihara et al. 2001). Excellent results were obtained in these studies due to the absence of cardiac motion, high-resolution, and wider field of view. Recently, fetal CMR has been expanded to much wider clinical cases with fetuses (Saleem 2008; Manganaro et al. 2008; Dong et al. 2013). Promising outcomes out of these studies have confirmed that fetal CMR technology is applicable in the assessment of fetal cardiac structure and function with exceptional image quality. However, some natural drawbacks, such as spatial and temporal resolution problem, as well as motion artifacts due to fetal cardiac movement and fetal motoric movement, occur hampering important information of the CMR (Sklansky 2004; Prakash et al. 2010). Still, the role of fetal echocardiography as the gold standard in prenatal detection of cardiac malformation cannot be replaced. The fetal CMR remains as a complementary by evaluating abnormalities and underlying etiologies that cannot be readily performed using fetal echocardiography (Frates et al. 2004).

3.2.2 Fetal Echocardiography: A State-of-the-Art Appraisal

Enormous progression in fetal echocardiography technology nowadays has enabled health care professionals to obtain more interactive and high-resolution detailed images of the growing fetus. This evolution also brings more possibilities to diagnose a much wider variety of structural malformations, genetic syndromes, and diseases from early stage of pregnancy (DeVore and Sklansky 2003; DeVore 2010; Mondillo et al. 2011; Li et al. 2013; Goncalves et al. 2004; Yagel et al. 2011). Furthermore, the advancement of ultrasound device for fetal echocardiography has triggered new varieties of applications, such as fetal assessment methods in multidimensional scheme (ranging from conventional two-dimensional/2-D, three-dimensional/3-D, and four-dimensional/4-D modes), multimodality imaging system, and fetal cardiovascular monitoring for fetoscopic surgery (Uittenbogaard et al. 2008; Elmstedt et al. 2011; Paladini et al. 2004; Kohl 2002). Beyond doubt, the involvement of imaging technology, image processing and analysis, and visualization techniques, combined with high-end ultrasound transducer system, is significant in developing such fetal echocardiography applications.

A large number of reviews on fetal echocardiography have been published over the last decade. Based on our findings, most of these reviews are blended between clinical and engineering fields since fetal echocardiography theme converges interdisciplinary collaborations. However, it can be intricate for researchers to track one sub-topic for their scientific purposes. In this regard, we intentionally categorize publications on fetal echocardiography reviews into two clusters, clinical and technical reviews, to streamline the topic tracking. The clinical reviews mostly emphasized on the accuracy and applicability of fetal echocardiography to detect and diagnose fetal cardiac anomalies, while the technical reviews generally underscored on the fetal echocardiography system and imaging techniques.

The review of Gembruch (1997) revealed the risk factors and the importance of taking punctual gestational age for the examination and improving personnel expertise in obtaining an accurate result. The importance of scanning in the exact gestational age and operator capability in improving the fetal echocardiography accuracy is also strengthened by Uittenbogaard et al. (2003) and Rasiah et al. (2006). The study of Gardiner (2001) and Comas et al. (2012) highlighted fetal cardiac function evaluation and its technical considerations by using current imaging modalities. This functional echocardiography is found to be important in selecting high-risk populations and several fetal conditions including intrauterine growth restriction, twin-to-twin transfusion syndrome, maternal diabetes, and congenital diaphragmatic hernia. M-mode annular displacement, pre-cordial venous Doppler flow assessment, and myocardial performance index (MPI) are considered to be the recommended methods. However, these modalities are perhaps most appropriate when pathological information is provided. Ayres (1997) described advances in transducer technology for fetal echocardiography. With support from computer processing and visualization, multidimensional acquisition can be performed dramatically. The implementation of new Doppler tissue imaging in combination with image resolution is also possible to provide obstetricians and pediatric cardiologists with more tools and techniques for earlier and more precise detection of fetuses with cardiac defects. Orwat et al. (2013) depicted the comparison of imaging modalities and their applications to diagnose adult CHD. Among the compared modalities, echocardiography noticeably remains the routine main imaging technique. The review of Li et al. (2013) discovered that three section views (four-chamber view, outflow tract view, and three-vessel trachea view) should be included in scan protocol to obtain great diagnostic potential for fetal echocardiography. Furthermore, extended cardiac echography examination (ECEE) as a specific protocol is proven to be capable of identifying some minimal defects in utero and provide more detailed information on suspicious fetal heart. While spatiotemporal image correlation (STIC) technology can be used to provide more detailed information for local situation of defects, especially for fetal cardiac intervention planning.

In technical reviews on imaging aspects of fetal echocardiography, the appraisals of Sklansky et al. (2004), Deng and Rodeck (2004), and Budorick and Millman (2000) seem to focus their studies on comparison of several imaging modalities for fetal cardiac examination and overview of supporting equipment and applications on fetal echocardiography. Budorick and Millman (2000) provided a comprehensive review on the fetal cardiac imaging modalities. The topic about 3-D ultrasound system was also explained thoroughly, ranging from gating system, real-time system, to the quantitative measurement of the scanned fetal heart. The new 3-D ultrasound system is shown to be capable of facilitating the evaluation of the dynamic function of the fetal heart for better analysis of complex cardiac anomalies in the future.

Out of the aforementioned technical reviews, we also noted several technical reviews that specifically concentrate on image processing, analysis, and visualization. Although the review of Whittingham (2007) outlined the physics of ultrasound imaging system, some of the analysis also focused on image acquisition scheme. In this regard, different modes of diagnostic scanning and techniques for improving image quality within the constraints of real-time operation have been remarked. Furthermore, the detailed review of Sklansky (2004) that underlined the 3-D ultrasound system for fetal echocardiography obviously provided brief overview about image acquisition, gating, processing, and display aspects. Not only 3-D ultrasound system, improvement in image resolution and screening techniques has also been discussed. The review of Deng and Rodeck (2004) has accentuated the use of image processing, analysis, and visualization in fetal cardiac imaging technology. A number of applications, such as dynamic three-dimensional echocardiography, myocardial Doppler imaging, harmonic ultrasound imaging, and B-flow sonography, have been summarized in terms of technical principles and clinical potentials. Moreover, the use of biomicroscopy, MRI, and multimodality imaging system has also been investigated to see the viability of such modalities for fetal cardiac imaging. Appropriate use and couse of these new tools are found not only to provide unique information for better clinical assessment of fetal cardiac disease, but also to offer new ways to improve understanding of cardiovascular development and pathogenesis. After all, it can be wrapped up that particular attention of the accessible reviews above is nowadays directed toward multidimensional imaging, specifically 3-D echocardiography, as a result of its comprehensive information in the prenatal diagnosis of CHD.

3.3 Basic Clinical Routine and Standards in Fetal Echocardiography

The development of the fetal heart begins at conception and completely formed by 8 weeks into pregnancy. Accordingly, CHD occurs during this development phase. Since the anatomy and physiology of the fetal heart is different from those of pediatric or adult, the abnormality is also unlike. In comparison with pediatric heart, the extension of fetoplacental circulation brings about difference in the assessment of the function in the fetal heart. Furthermore, due to extracardiac abnormalities and chromosomal defects, specific lesions may differ in characteristics. The fetal heart has specific abnormality spectrum, cardiac anatomy complexity, specific positioning, small size, and circulation differences which are represented by placental circulation of the blood to the fetal heart and return back to the placenta. Based on these characteristics, the requirement of an ideal fetal echocardiography for CHD imaging should be able to define all anatomical aspects of cardiac structure and evaluate physiological consequences of CHD. Therefore, performing an optimal fetal echocardiographic screening requires special professional prenatal sonographer with cardiac anatomy knowledge and suitable echocardiography device that generates high-resolution real-time images. Thanks to the development of the sophisticated imaging tool, detailed information on the fetal heart structure, function, and time-related events has become available (Gardiner 2001; Prakash et al. 2010; Lee et al. 2008).

The Clinical Standards Committee (CSC) of International Society of Ultrasound in Obstetrics and Gynecology (ISUOG) has a concern to develop practice guidelines and consensus statements that provide health care practitioners with a consensus-based approach for diagnostic imaging (Lee et al. 2008). The guidelines for performing basic and advanced procedures of fetal echocardiography have been largely documented and published in Lee et al. (2008), Barboza et al. (2002), Allan (2004). These procedures highlighted the scanning period based on gestational age, some technical factors for scanning, and cardiac examination in basic and extended methods (Lee et al. 2008). On the subject of our review on image processing perspective, we concentrate our discussion on scanning technical aspects.

Regarding the technical aspects for the examination, the echocardiography equipment for diagnostic purpose strongly depends on the application. Therefore, the choice of echocardiography application differs technically according to clinical constrains, such as maternal and fetal physical status, type of CHD diagnosis that needs to be obtained, gestational age, region of interest to be scanned, and so on that have been clearly revealed in the fetal echocardiography guidelines (Lee 2013) as well as technical constrains (type of transducer, technical settings, supporting equipment, acquisition and recording technique, processing system, data dimension visualization, and other technical requirements). However, we restrict our review only on the configuration for the aforementioned applications.

3.4 Imaging and Image Processing in Two-Dimensional Fetal Echocardiography

Rapid advancement in transducer technology and computer system has transmuted the 2-D imaging perspective into higher dimension of imaging system. Although the use of the 2-D echocardiography is considered to be old-fashioned, such scheme still becomes the gold standard for prenatal imaging of the fetal heart and situs (Rychik 2004). The important discussion in 2-D fetal echocardiography can be categorized into two groups: clinical and technical issues. Clinical issues encompass two points: challenges in performing diagnostic screening and the choice of transducers, while technical issues cover three aspects: ultrasound devices, acquisition techniques, and post-processing methods.

In clinical practice, 2-D fetal echocardiography is still powerful in clinical routine and plays a main role in obstetric diagnostics, especially when the availability of sophisticated multidimensional imaging modalities is limited. In diagnostic purpose, the so-called grayscale imaging scheme is the basis of a reliable fetal cardiac examination. In this scheme, 2-D fetal echocardiography is particularly used to observe the 'basic view' and 'extended basic view' screening for the four-chamber view, which allows assessment of abnormalities involving the atria and the ventricles, as well as right and left ventricular outflow tract and great artery, for more effective screening for CHD. However, many anomalies can still be missed (Gardiner 2001; Jaeggi et al. 2001; Lee 2013; Barboza et al. 2002; Allan 2004; Rychik 2004). Then, again, some clinical drawbacks on the 2-D-based method may occur hampering the scanning reliability. Occasionally, despite the regular 2-D fetal echocardiography procedure yields no abnormality results, particular undetected cardiac anomalies are still found sometime after birth (Sklansky 2004; Buskens et al. 1996). Additionally, separated assessment of the fetal heart and circulation may cause incomprehensive observation. This occurs because the pathophysiology of pregnancy to study uteroplacental circulation is performed by obstetrician using Doppler, while the morphological aspect of cardiac development is investigated by cardiologist by means of M-mode and Doppler (Gardiner 2001; Campbell et al. 1983; Allan et al. 1982, 1987). On the subject of the aforesaid clinical drawbacks, some factors have been thought to contribute to such problems, such as tremendously operator dependent with unreliable sonographic window. This limitation may be part of the cause of the commonly time-consuming acquisition.

- **Transducers and Acquisition Settings for 2-D fetal echocardiography**

On the choice of transducer, the accuracy of a fetal echocardiogram is affected by the scanning time during gestation when the study is performed and the transducer type to obtain the correct fetal heart characteristic. Basically, transvaginal echocardiography and transabdominal echocardiography are the two standard modalities currently being used in prenatal screening for congenital cardiac and extracardiac defects. Both transducers have been fostered primarily by the introduction of higher-frequency and higher-resolution ultrasound probes, so that they suit with early screening. However, transvaginal echocardiography is more applicable for observing cardiac defects at earlier gestational age of pregnancy (Gardiner 2001; Ayres 1997; Chitty and Pandya 1997; Budorick and Millman 2000).

In technical point of view, selecting the right transducer for certain application is indispensable to generate a correct and reliable image. In this regard, we draw our attention to define the appropriate transducer type, settings, and applications. The basic requirement of the 2-D fetal echocardiography transducer is excellent B mode, with a good cine-loop facility for scrolling back frame by frame and capturing the frame of interest as well as real-time scanning capability. Afterward, a special preset of a transducer for evaluating fetal heart is high frame rate, decreased persistence, and increased compression. The high frame rate is achieved by parallel processing, where the transducer transmits one line and receives two, resulting in a doubling of previous rates. The standardized frame rate setting for fetal echocardiography is 4–5 MHz. When the frame rate is decreased, the spatial resolution will be compensated as well (Gardiner 2001; Turan et al. 2009). Furthermore, the transducer frequency is the other feature that plays an important role in ultrasound image quality, as the higher the frequency, the higher the resolution and the greater anatomical detail can be obtained. While for the lower frequency, it is related to increased penetration of the sound beam (Sklansky 2004). In resolution setting, high-spatial, high-temporal, and high-contrast resolutions guarantee the accuracy of the diagnostic of most cardiac anomalies. The spatial resolution determines the anatomical details of the cardiac structures, while the temporal resolution reveals the motion factor (Turan et al. 2009; Chaubal and Chaubal 2009; Ng and Swanevelder 2011). For contrast resolution, it can be achieved by setting the dynamic range into low value. Additionally, the use of contrast agent is found to be effective in generating better contrast (Senior et al. 2009). However, it is considered to be unwise for the use of contrast agent in fetal echocardiography. The existence of artifacts is also a common phenomenon that determines the quality and diagnostic value of the image. Definitely, it is also necessary to provide color Doppler, pulsed Doppler, and continuous-wave Doppler for blood flow visualization and quantification. The availability of advanced fetal echocardiography applications, such as STIC, tissue Doppler, and multiplanar imaging, is certainly more advantageous (Chaubal and Chaubal 2009).

Once the transducer settings have been adjusted optimally and the acquisition have been performed, the other important technical issue in 2-D fetal echocardiography is image processing for automatic and quantitative analysis. Besides transducer characteristics, image processing is another important aspect that dramatically influences the accuracy of the diagnosis. The main reasons of involving image processing in this field are due to the facts that the echocardiography image quality is relatively poor and the complication of fetal echocardiography that requires specific experienced expert to perform the acquisition and analyze the data appropriately. Therefore, image processing research for fetal echocardiography is developed to improve the image quality and provide assistance in analyzing the data. Here, we listed some image processing techniques that have been widely developed and implemented for 2-D fetal echocardiography. As notification, not all reviewed methods are specifically for fetal echocardiography. In spite of that, the viability of implementation to fetal echocardiography is still highly relevant. However, further investigation is certainly demanded.

- **2-D fetal echocardiography speckle reduction and enhancement**

As a naturally multiplicative noise may reduce the image contrast, obscure edges and details, and hamper tissue structure visibility of ultrasound images, speckle problems necessitate to be clearly identified and accurately solved (Burckhardt 1978; Goodman 1976; Michailovich and Tannenbaum 2006; Wagner et al. 1983; Xie et al. 2002). The so-called speckle reduction technique is commonly utilized in ultrasound field as the prelude in ultrasound feature visualization enhancement (Zhu et al. 2009; Finn et al. 2011; Chen et al. 2003; Czerwinski et al. 1995; Hao et al. 1999; Tay et al. 2006). Besides visualization enhancement, speckle reduction may also function as image simplification where the complexity of the area of interest is reduced in such a way that it indirectly functions as image segmentation in the pre-feature extraction process (Yue et al. 2006; Rui et al. 2008).

In ultrasound applications, speckle reduction techniques can be divided into two main approaches: compounding and post-processing (Adam et al. 2006; Mateo and Fernandez-Caballero 2009). The compounding method reduces the speckle during ultrasound acquisition (Behar et al. 2003; Jespersen et al. 1998; Trahey et al. 1986), while the post-processing approach decreases such noise by employing filtering techniques after the acquisition. In general, the filtering techniques can be grouped into spatial (Chen et al. 2003; Czerwinski et al. 1995; Hao et al. 1999; Tay et al. 2006) and frequency domains (Zhou and Liu 2008; Cincotti et al. 2001).

Although the speckle reduction in general ultrasound has the same concepts as that of in echocardiography, we need to highlight some important issues about speckle in echocardiography, namely speckle noise location, type of speckle noise, and speckle characteristics. In terms of speckle noise location and characteristic, it is found that speckle noise in echocardiographic image is prominent in all cross-sectional views (Massay et al. 1989) and noticeably has more significant effect than additive noise sources (Zong et al. 1998). Furthermore, in most echocardiogram cases, endoborder usually has less information than epiborder information. Since differentiation of these borderlines has important information that represents cardiac disorder, it is important to characterize the speckle in the borderline (Choy and Jin 1998). However, noisy border information may affect border interpolation by human observers for the manually defined borders. Therefore, feature characterization and speckle reduction in the endo border area may be useful to improve the quality and visibility (Zong et al. 1998).

To address the echocardiography characteristics, some speckle reduction studies specifically for echocardiography have been developed. A review done by Finn et al. (2011) confirmed that most speckle reduction techniques have positively affected subjective image quality and improved boundary definition in the echocardiography images (Finn et al. 2011; Massay et al. 1989; Zong et al. 1998). Interestingly, from the comparison of several filtering techniques [Anisotropic Diffusion (Perona and Malik 1990; Yu and Acton 2002; Aja-Fernandez and Alberola-Lopez 2006; Weickert 1999; Abd-Elmoniem et al. 2002; Krissian et al. 2007), Wavelet denoising (Zong et al. 1998; Pižurica et al. 2003; Lee 1980; Frost et al. 1982; Kuan et al. 1987; Lopes et al. 1990), and Local Statistics (Crimmins 1985)], confirmed that anisotropic diffusion (AD), specifically oriented speckle reduction anisotropic diffusion (OSRAD) (Krissian et al. 2007) method presented the strongest speckle suppression and provided greatest average improvement in contrast relative to the unfiltered input.

With hindsight, the study about AD methods for reducing the speckle has been largely conducted for some periods. AD filters, as a part of the nonlinear diffusion filters, produce better preservation for edge strength and localization. The Perona-Malik filter enhances the classical AD filters by allowing parameter setting adjustment to improve image properties, especially edges (Perona and Malik 1990). The edge preservation was then improved by SRAD (Yu and Acton 2002). However, such filter cannot handle high-intensity speckles. The new variants of SRAD, such as generalized SRAD (Yu et al. 2004), OSRAD (Krissian et al. 2007), directions of gradient SRAD (Kim et al. 2008), multiscale-based adaptive SRAD

(Yoo et al. 2008), regularized SRAD (Yu and Yadegar 2006), detail-preserving anisotropic diffusion (DPAD) (Aja-Fernandez and Alberola-Lopez 2006), and adaptive window anisotropic diffusion (AWAD) (Liu et al. 2009), have generally shown enhanced performance over the classical one. Over again, OSRAD performs the best for echocardiography (Finn et al. 2011).

In the wavelet transform domain, Zhu et al. (2009), Finn et al. (2011), Mateo and Fernandez-Caballero (2009) performed comparative studies on speckle reduction for ultrasound images. As one kind of popular noise reduction methods, wavelet transform can be used for local analysis between time and frequency. With its multiscale analysis capability, it can extract the signal local singular characteristics. The focus capability in the filtering method contains the most of the signal that can be used for noise reduction (Zhu et al. 2009).

Based on the comparison of non-wavelet with some wavelet-based techniques for speckle reduction in ultrasound image (Zong et al. 1998; Thakur and Anand 2005; Hao et al. 1999), it seems that wavelet transform is not recommendable for ultrasound and echocardiography images (Zhu et al. 2009; Finn et al. 2011; Mateo and Fernandez-Caballero 2009). However, although the study of Finn et al. (2011) reveals that wavelet seems to be slightly unfavorable for echocardiography, the combination between nonlinear stretching and wavelet shrinkage techniques brings new advantage. It does not only reduce the speckle, but also enhance and restore features of interest, such as myocardial walls in 2-D echocardiograms from the parasternal short-axis view (Zong et al. 1998).

In real cases, image processing techniques sometimes still need to be improved and verified with clinical judgment to avoid incorrect speckle reduction. This is normal since clinicians do not need to remove the speckle in the actual manual diagnosis. Sometimes, clinicians have a preference to preserve the speckle, as it contains essential information. Therefore, the challenges in speckle reduction for 2-D echocardiography are mostly on determining the characteristics of the cardiac structures and the speckles in such locations, understanding which speckles need to be preserved and removed, and defining the best technique for optimum speckle reduction and feature preservation. In authors' point of view, speckle reduction techniques in 2-D fetal echocardiography are by and large comparable to those of in 2-D echocardiography. However, seeing that fetal heart has special features that differ from other populations, 2-D fetal echocardiography application needs to be adjusted to fit with the characteristics. In addition, for 3-D echocardiography application that employs 2-D conventional transducer, speckle reduction is usually favorably performed separately in this 2-D plane for the reason that the acquired speckle is in 2-D plane as well.

- **2-D fetal echocardiography segmentation**

In clinical practice, analysis and diagnosis of ultrasound images still rely on manual visual expertise. This labor-intensive technique is good, as direct au fait inspection may recognize specific cases that computer system cannot overcome yet. However, such process is considerably slow, has the tendency of inaccuracy due to overwhelming data, and may suffer from inconsistencies as well as observers' variability.

Therefore, the development of object localization and analysis techniques is helpful to help experts in diagnosis. As the first-line technique in cardiac diagnosis, quantitative assessment of cardiac structures using echocardiography becomes an important procedure for the expert to determine the abnormality. Seeing that size, shape, volume, and function of the cardiac structures provide essential parameters for the cardiac disorder analysis, image segmentation technique is an important aspect that should be available to generate such parameters.

Since image segmentation techniques have been developed from the early phase of classical image processing, a large number of methods have been investigated, implemented, and reviewed. Specifically for image segmentation in ultrasound images, an extensive review of Noble and Boukerroui (2006) could be a good reference. To address our preference, this review can give assistance by clustering the topics in image segmentation methods based on clinical application. For echocardiography application, the main issues for 2-D echocardiography lie in segmentation and tracking for endocardial borderline in the left ventricle (Noble and Boukerroui 2006). As contour of the borderline in the left ventricle is the primary concern, most of the studies are concentrated on determining the borderline accurately. Some studies in this area have been noted and classified by Noble and Boukerroui (2006).

Noble and Boukerroui (2006) highlighted several important segmentation methods for echocardiography images. Active contours (or snakes) and its combination with other methods have been done by Mishra et al. (2003), Mignotte and Meunier (2001), and Heitz et al. (1994). Bayesian framework for boundary segmentation was implemented by the studies of Mignotte et al. (2001) and Boukerroui et al. (2003), while level set was developed by Yan and Zhuang (2003), Lin et al. (2003). The work of Klinger et al. (1988) applied mathematical morphology technique for segmentation. At last, artificial neural network (ANN)-based methods have also been used for region-based segmentation in this application (Binder 2002; Friedland and Adam 1989).

Specifically for segmentation of fetal anatomic structures from echocardiographic images, Jardim and Figueiredo (2005) utilized contour estimation and observation model according to the maximum likelihood criterion via deterministic iterative algorithms. In this regard, contour estimation is formulated as a statistical estimation problem, where both the contour and the observation model parameters are unknown. The observation model (or likelihood function) relates, in probabilistic terms, the observed image with the underlying contour. This likelihood function is derived from a region-based statistical image model. The result of this study seems to be promising to estimate contours in a supervised manner, i.e., adapting to not completely known shapes and completely unknown observation parameters.

Then, again, the works on active contour have produced some studies on fetal cardiac segmentation. The study of Dindoyal et al. (2003) on deformable model tracking to segment and track ventricles in 2-D motion-gated fetal cardiac data employed a modified explicit gradient vector flow (GVF) snake with rigid body motion constraints combined with edge profile tracking using affine transformations to constrain the deformation. Furthermore, the study of Dindoyal et al. (2005) on level-set method for fetal heart chamber segmentation compared three level-set approaches: edge-based level set, region-based level set, and shape-based level set. The edge-based level set, which was introduced by Lassige et al. (2000) and based on the work of Yezzi et al. (1997) in the earlier stage, developed a level-set algorithm that is capable of segmenting septal defects (holes within the septal wall). The algorithm is known as edge-penalized constant advection (EPCA). The level set with region-based approach was based on the concept in Mumford-Shah (MS) energy functional for active regions (Chan and Vese 2001), edge flow of Sarti (2002), and level-set snake with region competition (Dindoyal et al. 2005). This approach generated an algorithm named MS Sarti Collision Detection (MSSCD). The shape-based level set, as in Dindoyal et al. (2005), essentially combines the MS driving force in Dindoyal et al. (2005) with exponential curvature dependence. This snake algorithm is referred to as template-initialized MS with shape prior term (TIMS + SP). The main driving forces in this snake are the shape prior and MS terms. The comparison of such three level-set approaches confirmed that EPCA snake gives best improvement with the addition of shape prior with manual tracing.

- **2-D fetal echocardiography analysis**

The role of experienced perinatal cardiologist is crucial in detecting or excluding significant fetal heart anomalies. In this regard, the use of fetal echocardiography has shown to be accurate to detect CHD. However, as the availability of the expert is limited and the diagnosis is also time-consuming, the use of image analysis to assist the CHD diagnosis is considered to be helpful. Thus far, the fetal heart evaluation using 2-D fetal echocardiography analysis requires measurement parameters, such as fetal heart position, cardiac axis, situs, structural delineation, and numerous characteristics of cardiac anomalies (Chaubal and Chaubal 2009). To simplify, we assume by ourselves that the 2-D fetal echocardiography analysis can be divided into three points, namely object localization, characterization, and measurement. The object localization is the boundary detection and segmentation as discussed above. The object characterization is more related to structural and textural analysis, while the object measurement is more to geometrical analysis.

In point of fact, the implementation of 2-D fetal echocardiography segmentation methods has largely contributed on analyzing the abnormalities in echocardiography image, specifically in quantifying geometrical parameters, such as area, volume, and motion of the cardiac structure. The study of Choy and Jin (1998) in the detection of endocardial borders has a way to quantify the characteristics of the cardiac structures. The studies of Jardim and Figueiredo (2005) and Dindoyal et al. (2003) that have been described above may also be included into the first stage of image analysis when the measurement of the segmented regions is calculated. As a result, the 2-D fetal echocardiography analysis has a propensity on image measurement, image analysis, and feature extraction techniques out of the segmented regions.

For object characterization, we draw our attention to texture analysis of the 2-D fetal echocardiography. As we see from the measurement parameters mentioned by Chaubal and Chaubal (2009), most of the characteristics of the fetal heart structure in the image are tissue structures. The review study of Kerut et al. (2003) on texture analysis in the characterization of myocardium tissue gives the impression about the importance of understanding the knowledge of tissue structure to define the best technique to characterize them. In this case, the physical aspects of interaction between the ultrasound wave and tissue need to be understood. To highlight ultrasound speckle, both diffuse scattering and Rayleigh scattering are essential factors in generating the visibility of tissue texture characteristics. If necessary, a verification of the echocardiography image with the actual human heart tissue by performing real incision may be helpful in understanding the problem. Therefore, studies in feature verification in this topic seem to be limited for specific purposes. However, we only limit our review on texture analysis for 2-D fetal echocardiography. Afresh, the study of Kerut et al. (2003) presented a nomenclature of texture analysis methods for echocardiography. Although no exact experimental result is shown, this study has given emphasizes about basic principle of several texture analysis and possible methods in classical texture analysis that can be investigated and analyzed to see the feasibility for implementation in 2-D fetal echocardiography.

The study of Deng et al. (2010) proposed a computerized method for the automated detection of fetal cardiac structure in the four-chamber-utilized active appearance model (AAM) for determining fetal cardiac structure after performing despeckling process. As a model that combines a shape model with a texture model, AAM (Cootes et al. 2001) built a shape model by aligning training data information and calculating statistical characteristics. Training data are generated by annotating images with labeled points as the key features or structures that represent the object. An eigenanalysis is used to perform a texture model. The active appearance model is achieved through learning the correlations of the shape and texture model. With this model, the shape and texture of the fetal heart can be efficiently obtained. However, the accuracy of this method strongly depends on preprocessing, which are despeckling, object segmentation, and image labeling.

Although the works of Carneiro et al. (2007, 2008) are not specifically for 2-D fetal echocardiography only, this study is more aligned with the state-of-the-art detection and top-down segmentation methods proposed in computer vision and machine learning. However, this work can be a good resource for comparative study of fetal anatomical tissue structures, including the fetal heart. In addition to the 2-D fetal echocardiography, new challenges come up, such as extreme appearance variability of the fetal abdomen and fetal body imaging, generalization to the same basic detection algorithm to all anatomical structures, and extreme efficiency. The classifier used for the structure detection is derived from the probabilistic boosting tree classifier (PBT). It is a boosting classifier where the strong classifiers are represented by the nodes of a binary tree. The results show high accuracy is clinically close to the accuracy of sonographers.

Out of the above-mentioned methods, there are still a large number of methods that may contribute to structure characterization and analysis of the 2-D fetal echocardiography image. However, at least the reviewed methods have represented the actual problems, solutions, and directions in this topic.

3.5 Future Direction and Conclusions

As the gold standard in the prenatal diagnosis, 2-D fetal echocardiography has brought a fundamental step in early detection and management of CHD during pregnancy. In comparison with CMD, 2-D fetal echocardiography is considered to be a few steps behind. However, this system has many advantages that the cutting-edge imaging technology still cannot beat, especially with its comprehensive examination, noninvasive, and easily mobile. Yet, running a 2-D fetal echocardiography always requires experience and a systematic approach. Guidelines for training have been specifically formulated for qualified operators only in view of the fact that acceptable sensitivity and specificity of the examination is operator dependent (Caserta et al. 2008).

Technically, the basic issues in imaging and image processing for 2-D echocardiography lie in three methods: speckle reduction, image segmentation, and image analysis. Based on the review, several techniques have been found to be robust, accurate, and effective to solve imaging problems. Besides the above-mentioned techniques, some other techniques may also contribute in improving the accuracy of the diagnosis, for instance, motion compensation, feature extraction and quantification, image classification, image visualization, image management and integration, and so on. With the involvement of image processing and analysis techniques in the 2-D fetal echocardiography system, new modifications, findings, and evaluation can be performed and bring more benefits in the obstetric care.

On the other hand, some technical problems, such as low image resolution and speckle noise that hamper a correct visualization, limited sonographic dimension and visualization that disable comprehensive multidimensional visibility, inadequate post-processing potential, and uncaptured dynamic movement of the fetal heart during beating due to still-image inadequate capacity, have a big potency to impede the accuracy of the 2-D fetal echocardiography examination. Therefore, the need of improved techniques in cardiac examination and new-generation ultrasound imaging system are vital in improving the diagnostic accuracy during prenatal cardiac monitoring.

Regarding the advancement of transducers for 2-D fetal echocardiography, it gives the impression that the development tendencies are more to generate acquisition system in higher dimension, in designs, electronics, computer architecture, and algorithms. To this end, 3-D imaging may be a solution. The 3-D fetal echocardiography research has been increasingly reported over the last decade (Deng et al. 1996; Acar et al. 2005; Hata et al. 2008). The 3-D echocardiography technology is evolving into real-time three-dimensional echocardiography (RT3DE) system. It ranges from a sparse array matrix transducer (2.5 or 3.5 MHz) consisting of 256 non-simultaneously firing elements to full matrix-array transducer ($\times 4$) generating ultrasonic beams in a phased array manner (Wang et al. 2003). Recently, a complex high-quality 3-D ultrasound system with fully connected 2-D matrix arrays having several thousand elements has been developed (Correale et al. 2008), whereas the freehand 3-D ultrasound system that modifies the 2-D matrix array transducer with positional acquisition system and post-processing system remain favorable due to its more economical and flexibility reasons.

Compared to 2-D, the 3-D fetal echocardiography has larger advantages. Since the volumetric data could be acquired in one sweep, this 3-D imaging system may facilitate visualization with reduced scanning time, operator dependence, and window dependence. With better view, the 3-D system may improve evaluation, interpretation, and understanding of the object of interest more interactively, as the visualization can be made in many options. Furthermore, thanks to comprehensive acquisition, the volumetric quantification of the object can be performed more accurately and reproducibly. Then, again, the higher the dimension of the data, the more complicated the system that needs to be taken into account. In developing 3-D fetal echocardiography system from acquisition to process and display, we will face some challenges in some topics as follows: 3-D transducer system, reconstruction, spatial compounding, registration, segmentation, feature extraction, classification, fusion, visualization, and so on. Some problems, such as slightly poor image resolution and motion artifacts, could be new prospective tasks for research.

Out of the 3-D concept, nowadays, the technology has moved fast to the 3-D/4-D system. The fetal echocardiography technology is also buzzing with the conception of 3-D/4-D ultrasound system in research and clinical practices. Although it is not widely applicable yet, it can become the standard of care for fetal echocardiography, on account of its potency to significantly improve the evaluation of the fetal heart. Newer applications, for instance, tomographic ultrasound imaging, spatiotemporal image correlation, inversion mode, speckle tracking, tissue harmonic imaging, and velocity vector imaging, are now growing greatly providing interesting supplementary information in the diagnosis of the CHD.

References

Abd-Elmoniem K, Youssef A-B, Kadah Y (2002) Real-time speckle reduction and coherence enhancement in ultrasound imaging via nonlinear anisotropic diffusion. IEEE Trans Biomed Eng 49(9):997–1014

Acar P, Dulac Y, Taktak A, Abadir S (2005) Real-time three-dimensional fetal echocardiography using matrix probe. Prenat Diagn 25(5):370–375

Adam D, Beilin-Nissan S, Friedman Z, Behar V (2006) The combined effect of spatial compounding and nonlinear filtering on the speckle reduction in ultrasound images. Ultrasonics 44(2):166–181

Aja-Fernandez S, Alberola-Lopez C (2006) On the estimation of the coefficient of variation for anisotropic diffusion speckle filtering. IEEE Trans Image Process 15(9):2694–2701

Zhu C et al (2009) Speckle noise suppression techniques for ultrasound images. In: 2009 4th international conference on internet computing for science and engineering

Allan L (2004) Technique of fetal echocardiography. Pediatr Cardiol 25:223–233

Allan LD, Joseph MC, Boyd EG et al (1982) M-mode echocardiography in the developing human fetus. Br Heart J 47:573–583

Allan LD, Chita SK, Al-Ghazali WH et al (1987) Doppler echocardiographic evaluation of the normal human fetal heart. Br Heart J 57:528–533

Allan LD, Sharland GK, Milburn A, Lockhart SM, Groves AMM, Anderson RH, Cook AC, Fagg NLK (1994) Prospective diagnosis of 1,006 consecutive cases of congenital heart disease in the fetus. J Am Coll Cardiol 23:1452–1458

Ayres (1997) Advances in fetal echocardiography. Tex Heart Inst J 24:250–259

Baardman ME, Kerstjens-Frederikse WS, Corpeleijn E et al (2012) Combined adverse effects of maternal smoking and high body mass index on heart development in offspring: evidence for interaction? Heart 98:474–479

Barboza JM, Dajani NK, Glenn LG, Angtuaco RDMSTL (2002) Prenatal diagnosis of congenital cardiac anomalies: a practical approach using two basic views. Radiographics 22:1125–1138

Behar V, Adam D, Friedman Z (2003) A new method of spatial compounding imaging. Ultrasonics 41:377–384

Binder T (2002) Three-dimensional echocardiography—principles and promises. J Clin Basic Cardiol 5(2):149–152

Boukerroui D, Baskurt A, Noble JA, Basset O (2003) Segmentation of ultrasound images— multiresolution 2D and 3D algorithm based on global and local statistics. Pattern Recognit Lett 24(4–5):779–790

Budorick NE, Millman SL (2000) New modalities for imaging the fetal heart. Semin Perinatol 24(5):352–359

Burckhardt CB (1978) Speckle in ultrasound B-mode scans. IEEE Trans Sonics Ultrason 25(1):1–6

Buskens E, Grobbee DE, Frohn-Mulder IME, Stewart PA, Juttmann RE, Wladimiroff JW et al (1996) Efficacy of routine fetal ultrasound screening for congenital heart disease in normal pregnancy. Circulation 94:67–72

Campbell S, Griffin DR, Pearce JM et al (1983) New Doppler technique for assessing uteroplacental blood flow. Lancet 3:675–677

Carneiro G, Georgescu B, Good S, Comaniciu D (2007) Automatic fetal measurements in ultrasound using constrained probabilistic boosting tree. medical image computing and computer-assisted intervention—MICCAI 2007. Lect Notes Comput Sci 4792:571–579

Carneiro G, Georgescu B, Good S, Comaniciu D (2008) Detection of fetal anatomies from ultrasound images using a constrained probabilistic boosting tree. IEEE Trans Med Imaging 27(9):1342–1355

Carvalho JS (2002) Improving the effectiveness of routine prenatal screening for major congenital heart defects. Heart 88(4):387–391

Carvalho et al (2013) ISUOG practice guidelines (updated): sonographic screening examination of the fetal heart. Ultrasound Obstet Gynecol 41:348–359

Caserta L, Ruggeri Z, D'Emidio, Coco C, Cignini P, Girgenti A, Mangiafico L, Giorlandino C (2008) Two-dimensional fetal echocardiography: where we are. J Prenat Med 2(3):31–35

Chan TF, Vese LA (2001) Active contours without edges. IEEE Trans Image Process 10(2):266–277

Chaubal NG, Chaubal J (2009) Fetal Echocardiography. Indian J Radiol Imaging 19(1):60–68

Chen Y, Yin R, Flynn P, Broschat S (2003) Aggressive region growing for speckle reduction in ultrasound images. Pattern Recogn Lett 24(4–5):677–691

Chew C, Halliday JL, Riley MM, Penny DJ (2007) Population-based study of antenatal detection of congenital heart disease by ultrasound examination. Ultrasound Obstet Gynecol 29:619–624

Chitty LS, Pandya P (1997) Ultrasound screening for fetal abnormalities in the first trimester. Prenat Diagn 17(13):1269–1281

Choy MM, Jin JS (1998) Extracting endocardial borders from sequential echocardiographic images—using mathemafical morphology ond temporal lnformation to lmprove contour accuracy. IEEE Eng Med Biol Mag 17(1):116–121

Chung T (2000) Assessment of cardiovascular anatomy in patients with congenital heart disease by magnetic resonance imaging. Pediatr Cardiol 21:18–26

Cincotti G, Loi G, Pappalardo M (2001) Frequency decomposition and compounding of ultrasound medical images with wavelet packets. IEEE Trans Med Imaging 20(8):764–771

Comas M, Crispi F (2012) Assessment of fetal cardiac function using tissue Doppler techniques. Fetal Diagn Ther 32:30–38

Cootes TF, Edwards GJ, Taylor CJ (2001) Active appearance models. IEEE Trans Pattern Anal Mach Intell 23(6):681–685

Correale M, Ieva R, Di Biase M (2008) Real-time three-dimensional echocardiography: an update. Eur J Intern Med 19(4):241–248

Crimmins T (1985) Geometric filter for speckle reduction. Appl Opt 24:1438–1443

Czerwinski RN, Jones DL, O'Brien Jr WD (1995) Ultrasound speckle reduction by directional median filtering. In: Proceedings of the international conference on image process, vol 1, pp 358–361

Deng J, Rodeck CH (2004) New fetal cardiac imaging techniques. Prenat Diagn 24:1092–1103

Deng J, Gardener JE, Rodeck CH, Lees WR (1996) Fetal echocardiography in three and four dimensions. Ultrasound Med Biol 22(8):979–986

Deng Y, Wang Y, Cheng P (2010) Automated detection of fetal cardiac structure from first-trimester ultrasound sequences. In: 3rd international conference on biomedical engineering and informatics (BMEI), vol 1, pp 127–131

DeVore GR (2010) Genetic sonography: the historical and clinical role of fetal echocardiography. Ultrasound Obstet Gynecol 35:509–521

DeVore Falkensammer, Sklansky Platt (2003) Spatio-temporal image correlation (STIC): new technology for evaluation of the fetal heart. Ultrasound Obstet Gynecol 22:380–387

Dindoyal I, Lambrou T, Deng J, Ruff CF, Linney AD, Todd-Pokròpek A (2003) An active contour model to segment foetal cardiac ultrasound data. Medical image understanding and analysis. University of Sheffield, UK, pp 77–80

Dindoyal I, Lambrou T, Deng J, Ruff CF, Linney AD, Rodeck CH, Todd-Pokropek A (2005) Level set segmentation of the fetal heart. Functional imaging and modeling of the heart. Lecture notes in computer science, vol 3504, pp 123–132

Dong SZ, Zhu M, Li F (2013) Preliminary experience with cardiovascular magnetic resonance in evaluation of fetal cardiovascular anomalies. J Cardiovasc Magn Reson 15:40

Elmstedt N, Lind B, Ferm-Widlund K, Westgren M, Brodin LA (2011) Temporal frequency requirements for tissue velocity imaging of the fetal heart. Ultrasound Obstet Gynecol 38:413–417

Fenton BW, Lin CS, Macedonia C, Schellinger D, Ascher S (2001) The fetus at term: in utero volume-selected proton MR spectroscopy with a breath-hold technique—a feasibility study. Radiology 219:563–566

Finn S, Glavin M, Jones E (2011) Echocardiographic speckle reduction comparison. IEEE Trans Ultrason Ferroelectr Freq Control 58(1):82–101

Frates M, Kumar A, Benson C, Ward V, Tempany C (2004) Fetal anomalies: comparison of MR imaging and US for diagnosis. Radiology 232:398–404

Friedland N, Adam D (1989) Automatic ventricular cavity boundary detection from sequential ultrasound images using simulated anneal. IEEE Trans Med Imaging 8(4):344–353

Frost VS, Stiles JA, Shanmugan KS, Holtzman JC (1982) A model for radar images and its application to adaptive digital filtering of multiplicative noise. IEEE Trans Pattern Anal Mach Intell 4(2):157–166

Gardiner HM (2001) Fetal echocardiography: 20 years of progress. Heart 86 (Suppl II):ii12–ii22

Garne E, Stoll C, Clement M, The Euroscan Group (2001) Evaluation of prenatal diagnosis of congenital heart diseases by ultrasound: experience from 20 European registries. Ultrasound Obstet Gynecol 17:386–391

Gembruch U (1997) Prenatal diagnosis of congenital heart disease. Prenat Diagn 17(13):1283–1298

Goncalves LF, Espinoza J, Lee W, Mazor M, Romero R (2004) Three- and four-dimensional reconstruction of the aortic and ductal arches using inversion mode: a new rendering algorithm for visualization of fluid-filled anatomical structures. Ultrasound Obstet Gynecol 24:696–698

Goodman JW (1976) Some fundamental properties of speckle. J Opt Soc Amer 66:1145–1150

Haak MC, van Vugt JMG (2003) Echocardiography in early pregnancy—review of literature. Ultrasound Med 22:271–280

Hao X, Gao S, Gao X (1999) A novel multiscale nonlinear thresholding method for ultrasonic speckle suppressing. IEEE Trans Med Imaging 18(9):787–794

Hata T et al (2008) Real-time three-dimensional color Doppler fetal echocardiographic features of congenital heart disease. J Obstet Gynaecol Res 34(4pt2):670–673

Heitz F, Perez P, Bouthemy P (1994) Multiscale minimization of global energy functions in some visual recovery problems. CVGIP Image Underst 59:125–134

Jaeggi ET, Sholler GF, Jones ODH, Cooper SG (2001) Comparative analysis of pattern, management and outcome of pre-versus postnatally diagnosed major congenital heart disease: a population-based study. Ultrasound Obstet Gynecol 17:380–385

Jardim SMGVB, Figueiredo MAT (2005) Segmentation of fetal ultrasound images. Ultrasound Med Biol 31(2):243–250

Jespersen SK, Wilhjelm JE, Sillesen H (1998) Multi-angle compound imaging. Ultrason Imaging 20(2):81–102

Kerut EK, Given M, Giles TD (2003) Review of methods for texture analysis of myocardium from echocardiographic images: a means of tissue characterization. Echocardiogr J CV Ultrasound Aliied Tech 20(8):727–736

Khattab K, Schmidheiny P, Wustmann K, Wahl A, Seiler C, Schwerzmann M (2013) Echocardiogram versus cardiac magnetic resonance imaging for assessing systolic function of subaortic right ventricle in adults with complete transposition of great arteries and previous atrial switch operation. Am J Cardiol 111(6):908–913

Kim HS, Yoon HS, Toan ND, Lee GS (2008) Anisotropic diffusion transform based on directions of edges. In: IEEE 8th international conference on computer and information technology workshops, pp 396–400

Klingler JWJ, Vaughan CL, Fraker TJ, Andrews LT (1988) Segmentation of echocardiographic images using mathematical morphology. IEEE Trans Biomed Eng 35(11):925–934

Kohl T (2002) Fetal echocardiography: new grounds to explore during fetal cardiac intervention. Surg Endosc Other Intervent Tech 23(3):334–346

Krissian K, Westin C-F, Kikinis R, Vosburgh K (2007) Oriented speckle reducing anisotropic diffusion. IEEE Trans Image Process 16(5):1412–1424

Kuan D, Sawchuk A, Strand T, Chavel P (1987) Adaptive restoration of images with speckle. IEEE Trans Acoust Speech Signal Process (ASSP) 35(3):373–383

Kurihara N, Tokieda K, Ikeda K, Mori K, Hokuto I, Nishimura O, Ishimoto H, Yuasa Y (2001) Prenatal MR findings in a case of aneurysm of the vein of Galen. Pediatr Radiol 31(3):160–162

Lagopoulos ME, Manlhiot C, McCrindle BW, Jaeggi ET, Friedberg MK, Nield LE (2010) Impact of prenatal diagnosis and anatomical subtype on outcome in double outlet right ventricle. Am Heart J 160(4):692–700

Lassige TA, Benkeser PJ, Fyfe D, Sharma S (2000) Comparison of septal defects in 2D and 3D echocardiography using active contour models. Comput Med Imaging Graph 24(6):377–388

Lee J-S (1980) Digital image enhancement and noise filtering by use of local statistics. IEEE Trans Pattern Anal Mach Intell PAMI-2(2):165–168

Lee W (2013) AIUM practice guideline—fetal echocardiography. American Institute of Ultrasound in Medicine

Lee W et al (2008) ISUOG consensus statement: what constitutes a fetal echocardiogram? Ultrasound Obstet Gynecol 32(2):239–242

Levi S (2002) Ultrasound in prenatal diagnosis: polemics around routine ultrasound screening for second trimester fetal malformations. Prenat Diagn 22(4):285–295

Li Y, Hua Y, Fang J, Wang C, Qiao L, Wan C, Mu D, Zhou K (2013) Performance of different scan protocols of fetal echocardiography in the diagnosis of fetal congenital heart disease: a systematic review and meta-analysis. PLoS One 8(6):e65484

Lin N, Yu WC, Duncan JS (2003) Combinative multi-scale level set framework for echocardiographic image segmentation. Med Image Anal 7(4):529–537

Liu G, Zeng X, Tian F, Li Z, Chaibou K (2009) Speckle reduction by adaptive window anisotropic diffusion. Sig Process 89(11):2233–2243

Lopes A, Touzi R, Nezry E (1990) Adaptive speckle filters and scene heterogeneity. IEEE Trans Geosci Rem Sens 28(6):992–1000

Lowry RB, Bedard T, Sibbald B, Harder JR, Trevenen C, Horobec V, Dyck JD (2013) Congenital heart defects and major structural noncardiac anomalies in Alberta, Canada, 1995–2002. Birth Defects Res 97:7986

Manganaro L, Savelli S, Di Maurizio M, Perrone A, Tesei J, Francioso A, Angeletti M, Coratella F, Irimia D, Fierro F, Ventriglia F, Ballesio L (2008) Potential role of fetal cardiac evaluation with magnetic resonance imaging: preliminary experience. Prenat Diagn 28:148–156

Massay RJ, Logan-Sinclair RB, Bamber JC, Gibson DG (1989) Quantitative effects of speckle reduction on cross sectional echocardiographic images. Br Heart J 62(4):298–304

Mateo JL, Fernandez-Caballero A (2009) Finding out general tendencies in speckle noise reduction in ultrasound images. Expert Syst Appl 36(4):7786–7797

Meyer-Wittkopf M, Cooper S, Sholler G (2001) Correlation between fetal cardiac diagnosis by obstetric and pediatric cardiologist sonographers and comparison with postnatal findings. Ultrasound Obstet Gynecol 17:392–397

Michailovich OV, Tannenbaum A (2006) Despeckling of medical ultrasound images. IEEE Trans Ultrason Ferroelectr Freq Control 53(1):64–78

Mignotte M, Meunier J (2001) A multiscale optimization approach for the dynamic contour-based boundary detection issue. Comput Med Imaging Graph 25(3):265–275

Mignotte M, Meunier J, Tardif J-C (2001) Endocardial boundary estimation and tracking in echocardiographic images using deformable template and markov random fields. Pattern Anal Appl 4(4):256–271

Mishra A, Dutta PK, Ghosh MK (2003) A GA based approach for boundary detection of left ventricle with echocardiographic image sequences. Image Vis Comput 21:967–976

Mondillo et al (2011) Speckle-tracking echocardiography a new technique for assessing myocardial function. J Ultrasound Med 30:71–83

Ng A, Swanevelder J (2011) Resolution in ultrasound imaging. Continuing Educ Anaesth Crit Care Pain 11(5):186–192

Noble JA, Boukerroui D (2006) Ultrasound image segmentation: a survey. IEEE Trans Med Imaging 25(8):987–1010

Orwat S, Diller GP, Baumgartner H (2013) Imaging of congenital heart disease in adults: choice of modalities. Eur Heart J Cardiovasc Imaging 15(1):6–17

Paladini D, Vassallo M, Tartaglione A, Lapadula C, Martinelli P (2004) The role of tissue harmonic imaging in fetal echocardiography. Ultrasound Obstet Gynecol 23:159–164

Pennell DJ (2010) Cardiovascular magnetic resonance. Circulation 121:692–705

Perona P, Malik J (1990) Scale-space and edge detection using anisotropic diffusion. IEEE Trans Pattern Anal Mach Intell 12(7):629–639

Pižurica A, Philips W, Lemahieu I, Acheroy M (2003) A versatile wavelet domain noise filtration technique for medical imaging. IEEE Trans Med Imaging 22(3):323–331

Prakash A, Powell AJ, Geva T (2010) Multimodality noninvasive imaging for assessment of congenital heart disease. Circ Cardiovasc Imaging 3:112–125

Rasiah SV, Publicover M, Ever AK, Khan KS, Kilby MD, Zamora J (2006) A systematic review of the accuracy of first-trimester ultrasound examination for detecting major congenital heart disease. Ultrasound Obstet Gynecol 28(1):110–116

Rui L, Zhuoxin S, Cishen Z (2008) Adaptive filter for speckle reduction with feature preservation in medical ultrasound images. In: IEEE international conference on control automation, robotics and vision, pp 1787–1792

Rychik J (2004) Fetal cardiovascular physiology. Pediatr Cardiol 25(3):201–209

Saleem SN (2008) Feasibility of magnetic resonance imaging (MRI) of the fetal heart using balanced steady-state-free-precession (SSFP) sequence along fetal body and cardiac planes. AJR 191:1208–1215

Sander TL, Klinkner DB, Tomita-Mitchell A, Mitchell ME (2006) Molecular and cellular basis of congenital heart disease. Pediatr Clin North Am 53:989–1009

Sarti A (2002) Subjective surfaces: a geometric model for boundary completion. Int J Comput Vis 46(3):201–221

Senior R, Becher H, Monaghan M, Agati L, Zamorano J, Vanoverschelde JL, Nihoyannopoulos P (2009) Contrast echocardiography: evidence-based recommendations by European Association of Echocardiography. Eur J Echocardiogr 10:194–212

Sklansky M (2004) Advances in fetal cardiac imaging. Pediatr Cardiol 25:307–321

Tay PC, Acton ST, Hossack JA (2006) Ultrasound despeckling using an adaptive window stochastic approach. In: Proceedings of the international conference on image process, pp 2549–2552

Thakur A, Anand RS (2005) Image quality based comparative evaluation of wavelet filters in ultrasound speckle reduction. Digit Signal Process 15(5):455–465

Trahey G, Smith S, van Ramm O (1986) Speckle pattern correlation with lateral aperture translation: experimental results and implications for spatial compounding. IEEE Trans Ultrason Ferroelectr Freq Control 33:257–264

Tsai-Goodman B, Geva T, Odegard KC, Sena LM, Powell AJ (2004) Clinical role, accuracy, and technical aspects of cardiovascular magnetic resonance imaging in infants. Am J Cardiol 94:69–74

Turan S, Turan OM, Ty-Torredes K, Harman CR, Baschat AA (2009) Standardization of the first-trimester fetal cardiac examination using spatiotemporal image correlation with tomographic ultrasound and color Doppler imaging. Ultrasound Obstet Gynecol 33(6):652–656

Uittenbogaard LB, Haak MC, Spreeuwenberg MD, van Vugt JMG (2008) A systematic analysis of the feasibility of four-dimensional ultrasound imaging using spatiotemporal image correlation in routine fetal echocardiography. Ultrasound Obstet Gynecol 31(6):625–632

Wagner RF, Smith SW, Sandrik JM, Lopez H (1983) Statistics of speckle in ultrasound B-scans. IEEE Trans Sonics Ultrason 30(3):156–163

Wang XF, Deng YB, Nanda NC, Deng J, Miller AP, Xie MX (2003) Live three-dimensional echocardiography: imaging principles and clinical application. Echocardiography 20(7): 593–604

Weickert J (1999) Coherence-enhancing diffusion filtering. Int J Comput Vis 31(2–3):111–127

Whittingham TA (2007) Medical diagnostic applications and sources. Prog Biophys Mol Biol 93:84–110

World Health Organization (2012) Congenital anomalies, fact sheet N°370. http://www.who.int/mediacentre/factsheets/fs370/en/

Xie H, Pierce LE, Ulaby FT (2002) Statistical properties of logarithmically transformed speckle. IEEE Trans Geosci Remote 40(3):721–727

Yagel S, Cohen SM, Rosenak D, Messing B, Lipschuetz M, Shen O, Valsky DV (2011) Added value of three-/four-dimensional ultrasound in offline analysis and diagnosis of congenital heart disease. Ultrasound Obstet Gynecol 37(4):432–437

Yan JY, Zhuang T (2003) Applying improved fast marching method to endocardial boundary detection in echocardiographic images. Pattern Recognit Lett 24(15):2777–2784

Yezzi A, Kichenassamy S, Kumar A, Olver P, Tannenbaun A (1997) A geometric snake model for segmentation of medical imagery. IEEE Trans Med Imaging 16(2):199–209

Yoo BC, Ryu JG, Park HK, Nishimura TH (2008) Multi-scale based adaptive SRAD for ultrasound images enhancement. In: Proceedings of the advances in electrical and electronics engineering, pp 258–266

Yu, Acton S (2002) Speckle reducing anisotropic diffusion. IEEE Trans Image Process 11(11):1260–1270

Yu Y, Yadegar J (2006) Regularized speckle reducing anisotropic diffusion for feature characterization. In: Proceedings of the international conference on image processing (ICIP), pp 1577–1580

Yu Y, Molloy JA, Acton ST (2004) Generalized speckle reducing anisotropic diffusion for ultrasound imagery. In: Proceedings of the IEEE symposium computer-based medical systems (CBMS), pp 279–284

Yue Y, Croitoru MM, Bidani A, Zwischenberger JB, Clark JW Jr (2006) Nonlinear multiscale wavelet diffusion for speckle suppression and edge enhancement in ultrasound images. IEEE Trans Med Imaging 25(3):297–311

Zhou X, Liu DC (2008) Interactive frequency compounding to medical ultrasound images. In: 7th Asian-Pacific conference on medical and biological engineering IFMBE proceedings, vol 19. Springer, Berlin, pp 286–289

Zong X, Laine AF, Geiser EA (1998) Speckle reduction and contrast enhancement of echocardiograms via multiscale nonlinear processing. IEEE Trans Med Imaging 17(4):532–540

Chapter 4
Texture-Based Statistical Detection and Discrimination of Some Respiratory Diseases Using Chest Radiograph

Norliza Mohd Noor, Omar Mohd Rijal, Ashari Yunus, Aziah Ahmad Mahayiddin, Chew Peng Gan, Ee Ling Ong and Syed Abdul Rahman Abu Bakar

Abstract This chapter proposes a novel texture-based statistical procedure to detect and discriminate lobar pneumonia, pulmonary tuberculosis (PTB), and lung cancer simultaneously using digitized chest radiographs. A modified principal component method applied to wavelet texture measures yielded feature vectors for the statistical discrimination procedure. The procedure initially discriminated between a particular disease and the normals. The maximum column sum energy texture measure yielded 98 % correct classification rates for all three diseases. The diseases were then compared pair-wise, and the combination of mean of energy and maximum value texture measures gave correct classification rates of 70, 97, and 79 % for pneumonia, PTB, and lung cancer, respectively.

Keywords Digital chest radiographs · Statistical discrimination · Pneumonia · Pulmonary tuberculosis · Lung cancer

N. Mohd Noor (✉)
Razak School of Engineering and Advanced Technology, Universiti Teknologi Malaysia,
UTM Kuala Lumpur Campus, Jalan Semarak, 54100 Kuala Lumpur, Malaysia
e-mail: norliza@utm.my

O. Mohd Rijal · C. P. Gan · E. L. Ong
Institute of Mathematical Science, University of Malaya, Lembah Pantai,
50603 Kuala Lumpur, Malaysia
e-mail: omarrija@um.edu.my

A. Yunus · A. A. Mahayiddin
Institute of Respiratory Medicine, Kuala Lumpur, Jalan Pahang 50586 Kuala Lumpur,
Malaysia
e-mail: ashdr64@yahoo.com.au

S. A. R. Abu Bakar
Faculty of Electrical Engineering, Universiti Teknologi Malaysia, 81310 Skudai-Johor,
Malaysia
e-mail: syed@fke.utm.my

K. W. Lai et al., *Advances in Medical Diagnostic Technology*,
Lecture Notes in Bioengineering, DOI: 10.1007/978-981-4585-72-9_4,
© Springer Science+Business Media Singapore 2014

4.1 Introduction

Tuberculosis and cancer are among the top killer diseases in the world where two million deaths worldwide are due to tuberculosis every year and about one million new cases of lung cancer have been detected annually (WHO 2003, 2006). In Malaysia, respiratory diseases accounted for 8.05 % of hospital admission with pneumonia being one of the top ten causes of death in Malaysian government hospital (Health Facts 2009). In 2008, the incidence rate of tuberculosis is 63.10 per 10,000 population and lung cancer incidence rate is 14.15 per 100,000 population for male and 6.1 per 100,000 for female (Health Facts 2009; Malaysian Cancer Statistics 2006). Lung cancer is third most common cancer among the population in Peninsular Malaysia (Malaysian Cancer Statistics 2006). The government's failure to curb smoking effectively contributes to high incidence of lung cancer especially in male population. The emergence of tuberculosis is partly due to the failure in diagnosing PTB patients seeking treatment for continuous cough (TB a Problem Once Again 2008). The similarity in terms of symptom and signs of the these three lung diseases makes diagnosing difficult for the medical practitioner.

Despite rapid advances in medical imaging technology, the conventional chest radiograph is still an important ingredient in the diagnosis of lung ailments (Ginnekien et al. 2001; Middlemiss 1982; Moores 1987). In Malaysia, government hospitals perform the diagnosis using radiograph films simply out of economic considerations.

The immediate problem with the use of chest X-rays concerns the use of considerable visual interpretation. Studies have shown that the accuracy of the interpretation is subject to varying degrees of observer error (Frieden 2004; Nakamura et al. 1970). This error includes the observer's inability to detect abnormal opacities and interpret them correctly, inter-observer variation (due to varying reading ability between observers) and intra-observer variation. The study done by Nakamura et al. (1970) stated that the observer error rates using radiograph images were high. Schillham et al. (2006) further confirmed that observer error still existed, and its rates were still high. Therefore, one of the important contribution of this study is the ability to reduce the error rates for detection. The need for computer-aided diagnosis (CAD), as a second opinion, for the medical practitioner is important in reducing the observer error.

In addition to the difficulty of using the conventional chest radiographs, there is an additional problem of the simultaneous discrimination for the three diseases. Related studies tend to address the problem of detecting and comparing a particular disease with normals. For example, Oliveira et al. (2007) studied the problem of pneumonia present and pneumonia absent using chest radiograph in detecting childhood pneumonia and van Ginneken et al. (2002) studied the problem of detecting pulmonary tuberculosis from mass TB screening program.

In using symptom and signs as the diagnosing tools, Hamilton et al. (2005) investigate clinical features used in detection of lung cancer and Gopi et al. (2007)

study the clinical features used in the diagnosis and treatment of tuberculous pleural effusion.

Simultaneous discrimination in the diagnosis of the three diseases is important when the prevalence of the three diseases is high. Effective and quick simultaneous screening will provide proper attention to the patient, and thus, the appropriate advanced test can be provided. This will eliminate the time-consuming diagnosis (eliminating disease one by one), currently practiced in most clinics. As far as this chapter is written, authors are not aware whether any studies have been done to discriminate the three diseases simultaneously.

A lot of work has been done on chest CAD algorithm for detection of lung nodules, interstitial opacities, cardiomegaly, vertebral fractures, interval changes in chest radiograph, classification of benign and malignant nodules, and the differential diagnosis of interstitial lung diseases (Schillham et al. 2006; Oliveira et al. 2007; van Ginneken et al. 2002; Hara et al. 2007).

Our proposed CAD algorithm is different from other semi-automatic methods in the sense that the selection of region of interest (ROI) does not involve the usual segmentation problem. The proposed statistical-based CAD algorithm does not depend on establishing precise boundaries. It is also not required to minimize any cost function associated with a given segmentation algorithm (Oliveira et al. 2007; van Ginneken et al. 2002; Hara et al. 2007; Katsuragawa and Doi 2007).

Similar studies in detecting lung abnormalities involving chest radiograph uses texture (Ginneken et al. 2002), contrast enhancement (Arzhaeva et al. 2009) and morphology features (Homma et al. 2009). Texture features in the form of moments of responses (standard deviation, skew, and kurtosis) extracted from multiscale filter banks for each ROI were considered in Ginneken et al. (2002). Their result showed a sensitivity of 0.86 at a specificity of 0.50 (area under the receiver operating curve is 0.820) in a TB mass screening program, which consist of 147 images with textural abnormalities and 241 were normal images, and a sensitivity of 0.97 at a specificity of 0.90 (area under the receiver operating curve is 0.986) when applied to a second database that consist of 100 normal images and 100 abnormal images. Arzhaeva et al. (2009) applied multiscale filter bank of Gaussian derivatives and obtained moments of histograms as texture features. The authors used dissimilarity-based classification, which resulted in an area of 0.98 under the receiver operating curve. Katsuragawa and Doi (2007) enhanced the image of lung nodules by a difference-image technique and hence applied a rule-based method as the classifier. Their result showed a recognition rate of 98.5 % for 1,681 cases, and even for 22 misfiled cases, 86.4 % were correctly identified. Homma et al. (2009) used morphological filters and achieved high true positive rate for their CAD system for detecting lung cancer using X-ray CT.

This study concentrates on developing the algorithm for feature extraction useful for differentiating lung diseases, which are very similar in clinical symptoms and sign, namely lobar pneumonia, pulmonary tuberculosis, and lung cancer. These features are the input to a novel discrimination procedure. The algorithm developed has been used to develop a semi-automated CAD system. It should be emphasized that the CAD was designed as a low cost system where the only imaging modality utilized is the chest X-ray image, hence, other imaging modality such as MRI and volumetric CT were not considered.

Error rates are defined for the pair-wise discrimination of two types of diseases. For example, when discriminating between two types of lung conditions A_j and A_k, the error rates are given as:

$$\alpha = P(\text{Type I Error}) = P(A_j|A_k)$$
$$\beta = P(\text{Type II Error}) = P(A_k|A_j)$$

where $j \neq k, j = 1, ..., 4, k = 1, ..., 4$.

The infected region or ROI cannot be easily represented by standard measurement of length, area, shape, and size causing the selection of feature vectors difficult for any discrimination procedure. Henceforth, any standard image processing technique such as image enhancement or segmentation is avoided as much as possible to avoid possible lost of information from the original image.

4.2 Materials and Methods

4.2.1 Selection of Case Study

This study involved collaboration with the Institute of Respiratory Medicine (IPR), Malaysia, which is the national referral center for respiratory diseases. Cases that arrived at the IPR may be considered a random sample since an individual case may come from any of the Malaysian hospitals or clinics. The IPR provided archived patients' data which include chest X-ray films captured using the Phillips Diagnost 55/Super 50CP (Phillips Corp., Holland) together with complete patients' medical information. The patients' chest were captured in full inspiration using the posterior–anterior (PA) view with distance from the X-ray to the patient is fixed at 180 cm to diminish the effect of beam divergence and magnification of structures closer to the X-ray tube. The cassette size of 35 × 35 cm is used for female chest and 35 × 34 cm for male chest. The patient is exposed to 64 kV and 4.0 mAs if underweight, and 70 kV and 5.0 mAs if the patient has normal weight.

The archived data (stored in files) in IPR were diagnosed by a pulmonologist. In IPR, all the pulmonologists are trained to interpret chest radiographs. Stratified random sampling (SRS) was carried out for the patients' file. SRS means that files were randomly selected given that the patients chosen were already diagnosed as PTB or LC. The role of the consultant pulmonologist is to verify the diagnosis. It should be noted that the pulmonologist and consultant pulmonologist mentioned above are two different individuals.

The patient's chest X-ray is then divided into two groups, which are the control group and the test group. The selected patients used as the control group were the confirmed pneumonia (PNEU), pulmonary tuberculosis (PTB), and lung cancer (LC) cases with no other systemic diseases such as diabetes, hypertension, and heart disease. The omission of cases with other systemic diseases was done in order to avoid bias in the development of the statistical discriminant function (DF). The test group was selected similarly except that some of the patients may have other systemic diseases.

Lobar pneumonia is defined when one section or lobe of the lung is affected. In diagnosing pneumonia, the patient is assessed for crackles and wheezing sound from the lung, using a stethoscope (Wipf et al. 1999). The radiographic interpretation is considered the gold standard for the presence of pneumonia where the physical findings is accurate if it is found in the same location as an infiltrate on chest X-ray (Wipf et al. 1999). The confirmation of the PTB cases is based on the clinical feature (symptoms and sign), chest X-ray examination, and sputum Acid Fast Bacilli (AFB) direct smear. For the lung cancer, cases under study consist of 75 % cases of non-small cell carcinoma (of which 50 % are the squamous cell carcinoma and 25 % are the adenocarcinoma) and 25 % cases of small cell carcinoma. The confirmation of LC was based on bronchial biopsy result. The normal lung (NL) chest X-ray films selected by the radiologist from Universiti Sains Malaysia Hospital (HUSM) represent patients who came for a general medical checkup.

Patients that have lung disease (either PNEU, PTB, or LC) will have their chest X-ray film image shows some abnormal opacity. The existence of lung consolidation in the chest X-ray may confirm the existence of pneumonia and may appear on the chest X-ray after a few days of infection. The PTB image will show multiple opacities of varying size that run together (coalesce) in the chest X-ray image. Severe cases of PTB may result in consolidation and cavity, and the scarring marks may remain visible in the chest X-ray even after the patient is cured. Lung cancer appears as a mass opacity in the chest X-ray image. The chest radiograph image of a normal lung will show a complete dark image between rib bones due to nonexistence of any hardened substances.

The chest X-ray films were then digitized into DICOM format using the Kodak LS 75 X-ray Film Scanner (pixel spot size of 100 μm, 12 bit per pixel, image size of 2016 × 2048 pixels). An example of a digitized X-ray film is shown in Fig. 4.1a–d.

4.2.2 Texture Measures

Each of the ROI for a given image was subjected to the two-dimensional Daubechies wavelet transform as shown in Fig. 4.2, (Daubechies 1992; Walker 1999). The wavelet transform convert the image into four subsets, labeled LL, LH, HL, and HH representing the trend, horizontal, vertical, and diagonal detail coefficients.

The twelve texture measures considered were as follows:

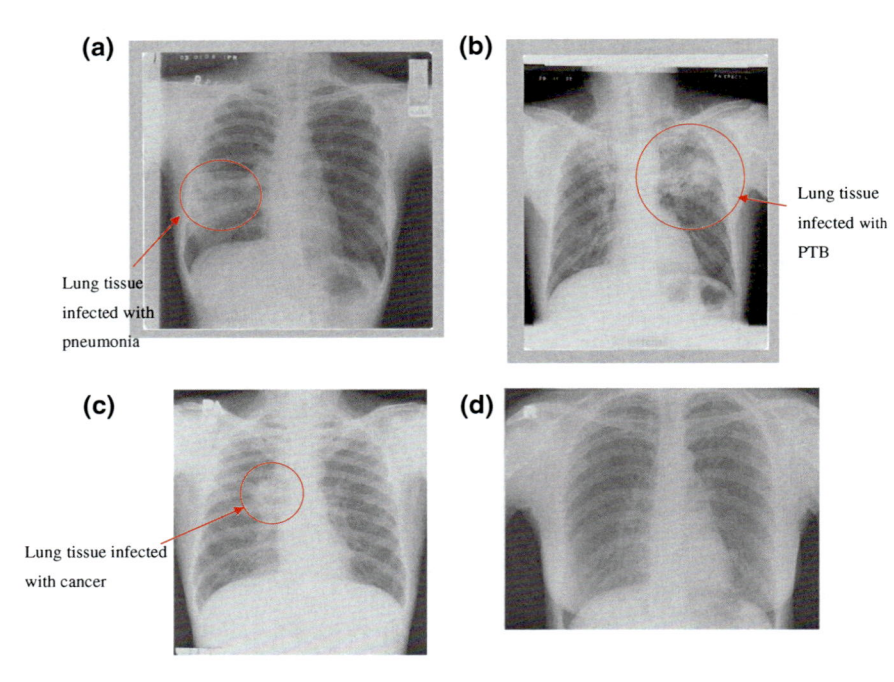

Fig. 4.1 a Visual of chest radiograph of pneumonia-infected lung (*source* The Institute of Respiratory Medicine, Kuala Lumpur). **b** Example of chest radiograph showing PTB-infected lung (snowflakes) (*source* The Institute of Respiratory Medicine, Kuala Lumpur). **c** Example of chest radiograph showing lung cancer (*source* The Institute of Respiratory Medicine, Kuala Lumpur). **d** Example of normal lung of an healthy individual (*source* The Institute of Respiratory Medicine, Kuala Lumpur)

1. Mean Energy, $E = \frac{1}{N} \sum_j \sum_k |C_{jk}|^2$
2. Entropy $= -\frac{1}{N^2} \sum_j \sum_k |C_{jk}|^2 \log |C_{jk}|^2$
3. Contrast $= \sum_j \sum_k (j-k)^2 C_{jk}$
4. Homogeneity, $H = \sum_j \sum_k \frac{C_{jk}}{1+|j-k|}$
5. Standard deviation of value, $\text{STDV} = \sqrt{\frac{1}{N^2} \sum_j \sum_k (C_{jk} - \mu)^2}$ where $\mu = \frac{1}{N^2} \sum_j \sum_k C_{jk}$
6. Standard deviation of energy, $\text{STDE} = \sqrt{\frac{1}{N^2} \sum_j \sum_k \left(|C_{jk}|^2 - \mu\right)^2}$ where $\mu = \frac{1}{N^2} \sum_j \sum_k |C_{jk}|^2$
7. Maximum wavelet coefficient value, $\max = \max(C_{jk})$
8. Minimum wavelet coefficient value, $\min = \min(C_{jk})$
9. Maximum value of energy, $E_{\max} = \max\left(\sum_j \sum_k |C_{jk}|^2\right)$
10. Maximum row sum energy

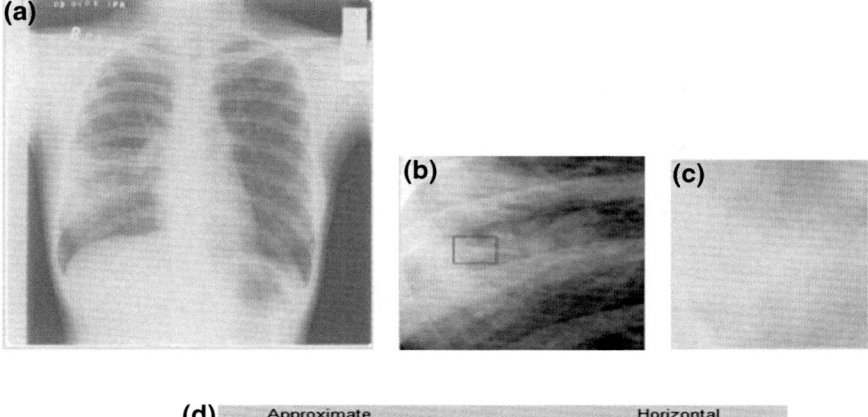

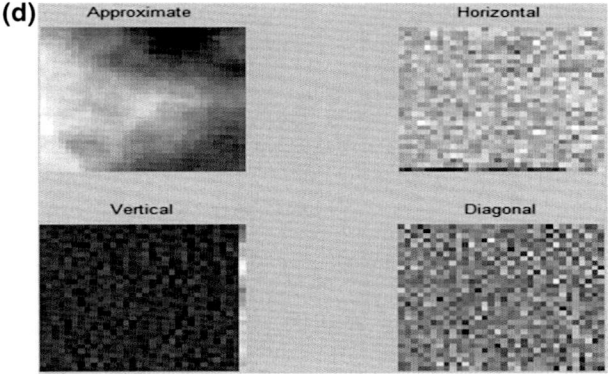

Fig. 4.2 a Chest X-ray of a pneumonia patient and **b** a subset image of the infected area. **c** Region of interest and **d** the transformed image where four image subset was formed (*source* The Institute of Respiratory Medicine, Kuala Lumpur)

11. Maximum column sum energy
12. Average number of zero-crossings

where C_{jk} is the element of sub-image (say, LL) found in row-j and column-k (Gonzalez and Woods 1992; Sonka et al. 1998). Hence, twelve texture measures in each of LL, LH, HL, and HH yield 48 descriptors or features, $\underline{u}$, that will be used to detect pneumonia.

4.2.3 Modified Principal Component Method

The modified principal component (ModPC) method was introduced in (Noor et al. 2010) where PNEU is discriminated from normals. The ModPC method is now extended for pair-wise comparison between three types of diseases namely, PNEU, PTB, LC, and normals.

A sample of 200 images were concurrently read and interpreted for the presence of PNEU, PTB, LC, and normals by two independent pulmonologists who are trained according to the World Health Organization (WHO) guideline (WHO Report 2004; Cherian et al. 2005), and the affected region (ROI) was identified.

The data used were divided into two sets, $(\underline{u}_1, \underline{u}_2, \ldots, \underline{u}_{120})$ as the control data set and $(\underline{u}_{121}, \underline{u}_{122}, \ldots, \underline{u}_{200})$ as the test data set. Let,

$\underline{u}_1, \underline{u}_2, \ldots, \underline{u}_{30}$ where $\underline{u} \in G_1$ represents the texture measures for PNEU samples, $\underline{u}_{31}, \underline{u}_{32}, \ldots, \underline{u}_{60}$ where $\underline{u} \in G_2$ represents the texture measures for normal lung samples, $\underline{u}_{61}, \underline{u}_{62}, \ldots, \underline{u}_{90}$ where $\underline{u} \in G_3$ represents the texture measures for PTB samples, and $\underline{u}_{91}, \underline{u}_{92}, \ldots, \underline{u}_{120}$ where $\underline{u} \in G_4$ represents the texture measures for LC samples.

The main problem of the ModPC method is the choice of an orthogonal transformation. Let M be an orthogonal matrix such that

$$\underline{u}_r^* = M\underline{u}_r \ (r = 1, \ldots, 200).$$

Let $\frac{1}{n-1}S_j$ be the estimate of the covariance matrix for group G_j ($j = 1, \ldots, 4$). For example,

$$S_1 = \sum_{r=1}^{30} (\underline{u}_r - \overline{u})(\underline{u}_r - \overline{u})^T \quad \text{where } \overline{u} = (\underline{u}_1 + \underline{u}_2 + \cdots + \underline{u}_{30})/30$$

while S_2, S_3, and S_4 were similarly defined for the sets G_2, G_3, and G_4, respectively.

The spectral decomposition of the estimated covariance matrices are

$$\frac{1}{n-1}S_j = Q_j \Lambda_j Q_j^T \quad \text{for } G_j \ (j = 1, 2, 3, 4)$$

where $n = 30$, $Q_j(j = 1, \ldots, 4)$ is the appropriate matrix of eigenvectors, and Λ_j ($j = 1, \ldots, 4$) is the corresponding diagonal matrix of eigenvalues. Henceforth, the choice of M will depend on minimizing misclassification probabilities in a two population discrimination problems. In particular, choose $M = Q_j$ or $M = Q_k$ ($j \neq k$), $j = 1, \ldots, 4$ and $k = 1, \ldots, 4$ such that the probability of misclassifying the test data to either population-j or population-k is minimized.

Without loss of generality, consider the two population discrimination problems PNEU and NL. For a selected M matrix, take the first two components of $\underline{u}_r^*(r = 1, \ldots, 30)$, which explain at least 90 % of the variability, relabel it as $\underline{v}_r$ ($r = 1, \ldots, 30$), and perform the following:

For vectors, $\underline{v}_1, \underline{v}_2, \ldots, \underline{v}_{30}, \in \Re^2$ calculate the statistics $\overline{v}_1 = (\underline{v}_1 + \cdots + \underline{v}_{30})/30$ and $S_{v1} = \sum_{j=1}^{30} \left(\underline{v}_j - \overline{v}_1 \right)\left(\underline{v}_j - \overline{v}_1 \right)^T$. The vectors $\underline{v}_1$, $\underline{v}_2$, ..., $\underline{v}_{30}$ were found to be bivariate normal (see Sect. 4.2.4). Henceforth, the PNEU ellipsoid $(\underline{v} - \overline{v}_1)^T((n-1)S_{v1}^{-1})(\underline{v} - \overline{v}_1) = c$ was drawn where c was selected from a standard

Chi square table (see Sect. 4.2.5). Further, the estimate of $g_1(\underline{v})$, which is the probability distribution for G_1 was also obtained.

The above was repeated for $\underline{v}_{31}, \underline{v}_{32}, \ldots, \underline{v}_{60}$ yielding, say the NL ellipsoid and the corresponding estimate of $g_2(\underline{v})$ which is the probability distribution for G_2. Finally, the estimate of the discriminant function $DF_{12}(\underline{v}) = \ln \frac{g_1(\underline{v})}{g_2(\underline{v})}$ may be derived (Johnson and Wichern 2007). Two-dimensional probability ellipsoids and appropriate DFs estimate the following error probability;

$$\alpha = P(\text{Type 1 Error}) = P(\text{PNEU}|\text{NL}) \tag{4.1}$$

and

$$\beta = P(\text{Type 2 Error}) = P(\text{NL}|\text{PNEU}) \tag{4.2}$$

for a selected texture measure.

For the PNEU–NL discrimination problem, there are two ways of estimating the error probabilities α and β by using the test set $\underline{v}_{121}, \underline{v}_{62}, \ldots, \underline{v}_{140}$ from G_1 and $\underline{v}_{141}, \underline{v}_{82}, \ldots, \underline{v}_{160}$ from G_2 in two ways where $\underline{v}_j$ $(j = 121, \ldots, 160)$ are the first two components of $\underline{u}_r^* = M\underline{u}_r = Q\underline{u}_r (r = 121, \ldots, 160)$;

(a) Estimation of α and β from the probability ellipsoid:

    1. The number of times $\underline{v}_j$ $(j = 121, \ldots, 140)$ falls into the NL ellipsoid gives an estimate of β.

    2. The number of times $\underline{v}_j$ $(j = 141, \ldots, 160)$ falls into the PNEU ellipsoid gives an estimate of α.

(b) Estimation of α and β from DF:

    Investigate if $DF_{12}(\underline{v}) = \ln \frac{g_1(\underline{v})}{g_2(\underline{v})} < \log K$

    where $g_1(\underline{v})$, the probability distribution for PNEU, was found to be $N_2(\underline{\mu}_1, \Sigma_1)$ and $g_2(\underline{v})$, the probability distribution for NL was shown to be $N_2(\underline{\mu}_2, \Sigma_2)$. Further, $K = \frac{d(1|2)}{d(2|1)} \frac{p_2}{p_1}$ where $d(i|j)$ is the cost of misclassifying observation-j ($i = 1, 2$ and $j = 1, 2$), while p_1 and p_2 are the a priori probabilities. Suppose $\underline{v}^*$ is an unknown observation and assuming that $p_1 = p_2$ and $d(1|2) = d(2|1)$, then $\underline{v}^*$ is assigned to the PNEU group if $DF_{12}(\underline{v}^*) > 0$, otherwise it is assigned to the NL group.

The equality of covariance matrices was tested using the Box's Test (Mardia et al. 1979), and if $\Sigma_1 = \Sigma_2$, $DF_{12}(\underline{v})$ is the linear discriminant function (LDF), which allocates the unknown observation $\underline{m}_0$ as follows;

Allocate $\underline{m}_0$ to population one if

$$\left[(\underline{\mu}_1 - \underline{\mu}_2)^T \Sigma^{-1}(\underline{m}_0) + \frac{1}{2}(\underline{\mu}_1 - \underline{\mu}_2)^T \Sigma^{-1}(\underline{\mu}_1 + \underline{\mu}_2) \right] \geq \ln\left[\left(\frac{d(1|2)}{d(2|1)} \right) \left(\frac{p_2}{p_1} \right) \right] \tag{4.3}$$

Table 4.1 Pair-wise discrimination strategy between disease absent (DA) and disease present (DP)

Pair-wise component	Type I error	Type II error
Pneumonia–normal lung (PNEU–NL)	P(PNEU|NL)	P(NL|PNEU)
Pulmonary tuberculosis–normal lung (PTB–NL)	P(PTB|NL)	P(NL|PTB)
Lung Cancer–normal lung (LC–NL)	P(LC|NL)	P(NL|LC)

Table 4.2 Pair-wise discrimination strategy between diseases

Pair-wise component	Type I error	Type II error
PNEU–PTB	P(PNEU|PTB)	P(PTB|PNEU)
PNEU–LC	P(PNEU|LC)	P(LC|PNEU)
PTB–LC	P(PTB|LC)	P(LC|PTB)

Otherwise allocate $\underline{m}_0$ to population two.

Alternatively, if $\Sigma_1 \neq \Sigma_2$ then $DF_{12}(\underline{v})$ is the quadratic discriminant function (QDF), which allocates the unknown $\underline{m}_0$ as follows;

Allocate $\underline{m}_0$ to population one if

$$\left[-\frac{1}{2}\underline{m}_0^T \Sigma_1^{-1} - \Sigma_2^{-1}\underline{m}_0 + (\underline{\mu}_1^T \Sigma_1^{-1} - \underline{\mu}_2^T \Sigma_2^{-1})^T \underline{m}_0 - k \right] \geq \left(\frac{d(1|2)}{d(2|1)} \right) \left(\frac{p_2}{p_1} \right) \quad (4.4)$$

where $k = \frac{1}{2}\ln\left(\frac{|\Sigma_1|}{|\Sigma_2|}\right) + \frac{1}{2}\left(\underline{\mu}_1^T \Sigma_1^{-1}\underline{\mu}_1 - \underline{\mu}_2^T \Sigma_2^{-1}\underline{\mu}_2 \right)$.

Otherwise allocate $\underline{m}_0$ to population two.

Throughout the study it is assumed that $d(1|2) = d(2|1)$ and $p_1 = p_2$ for both Eqs. 4.3 and 4.4. These assumptions were taken because the event of having either disease is regarded with having equal weight, and equal a prior probability is because there is no true or exact information about total frequency of cases in Malaysia.

Henceforth, the number of times $DF_{12}(\underline{v}_j) < 0$ for $\underline{v}_j$ $(j = 121, \ldots, 140)$ gives an estimate of β. Likewise, α is similarly derived.

All the above were repeated for the second selection of M (say $M = Q_k$) and suppose this choice yields lower α and β values, then $M = Q_k$ will be the preferred choice. Tables 4.1 and 4.2 illustrate all the combination of the two population discrimination problem studied. Flowchart shown in Fig. 4.3 illustrates the discrimination problem for the disease present and disease absent cases. Flowchart shown in Fig. 4.4 gives similar illustration for the pair-wise comparison of diseases.

4.2.4 Testing for Normality

Given $\underline{v}_1, \underline{v}_2, \ldots, \underline{v}_n \in \Re^2$. The statistics $\overline{\underline{v}} = (\underline{v}_1 + \cdots + \underline{v}_n)/n$ and $S_v = \sum_{j=1}^{n} (\underline{v}_j - \overline{\underline{v}})(\underline{v}_j - \overline{\underline{v}})^T$ and $y_j = (\underline{v}_j - \overline{\underline{v}})^T \left((n-1)S_v^{-1} \right)(\underline{v}_j - \overline{\underline{v}})$ for $j = 1, \ldots, n$ were

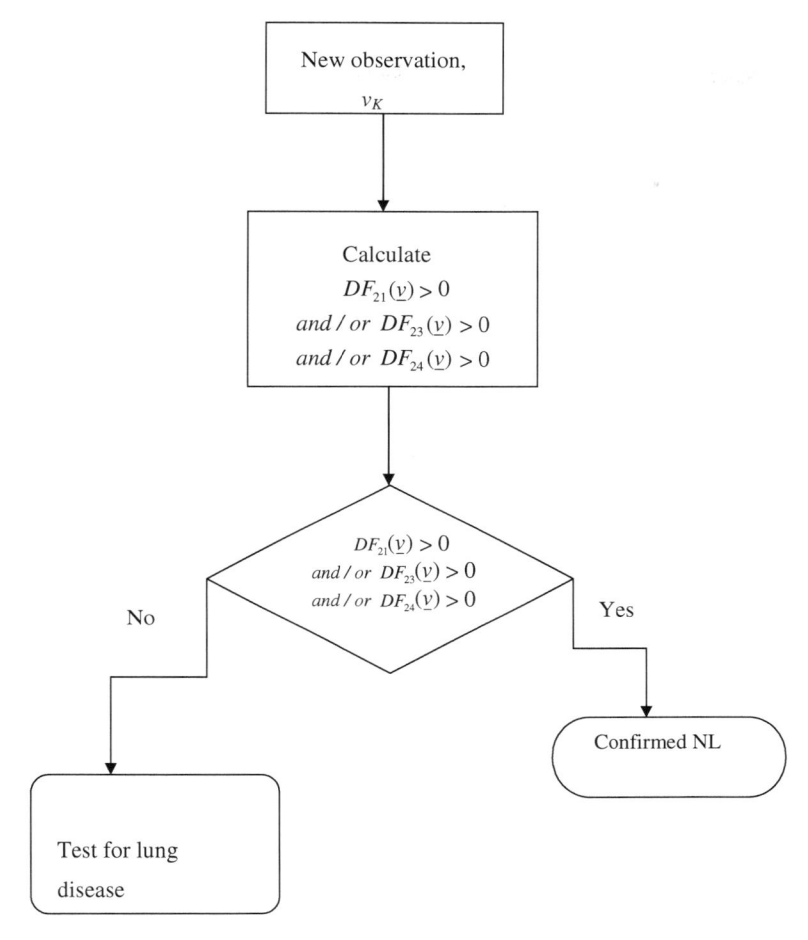

Fig. 4.3 The discrimination procedure flowchart for discriminating disease present and disease absent

calculated. If v_j is normally distributed, then y_j $(j = 1, \ldots, n)$ must come from a Chi squared distribution with two degrees of freedom. Effectively, testing normality of $v_1, v_2, \ldots, v_n$ is equivalent to testing whether y_j $(j = 1, \ldots, n)$ comes from a Chi squared distribution using the Kolgomorov–Smirnov test, (Mardia et al. 1979). In all cases studied, the Kolmogorov–Smirnov test confirms that y_j has a Chi squared distribution, henceforth, indicating that v_j is bivariate normal.

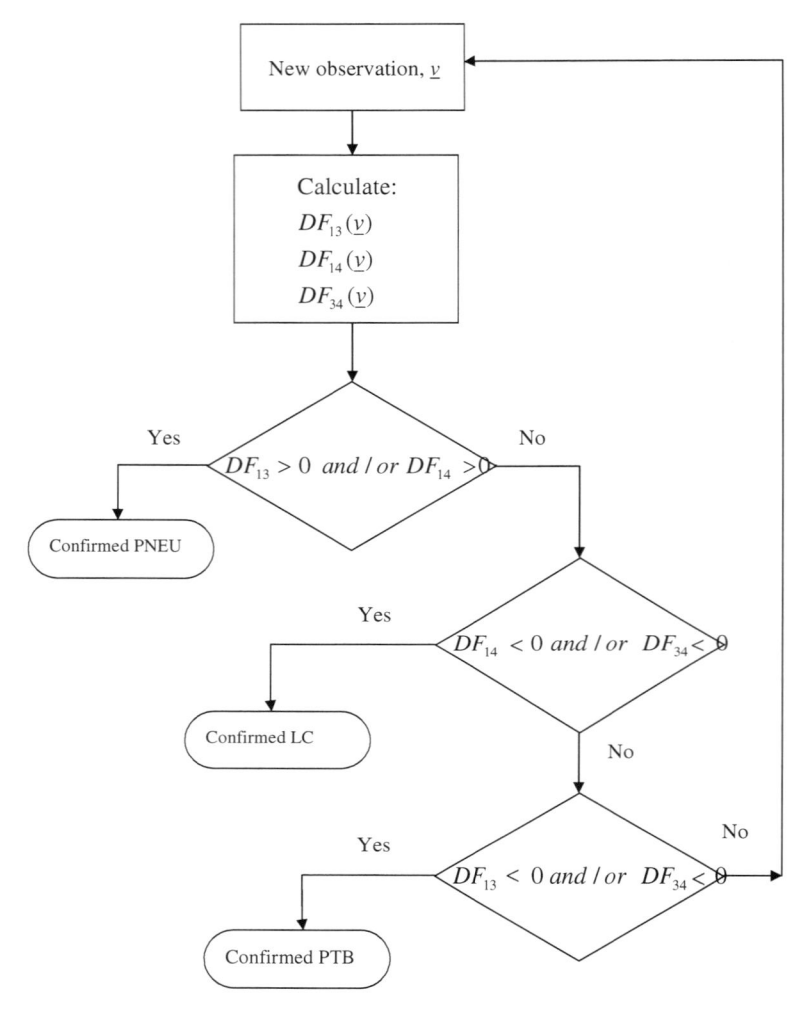

Fig. 4.4 The discrimination procedure flowchart for discriminating PNEU, PTB, and LC

4.2.5 Confidence Region of the First Two Components of $\underline{v}$

A 2D plot of the first two components of $\underline{v}$ together with its corresponding approximate confidence region (an ellipse) was used to investigate clustering of the two groups of individuals. If two clusters of points are discovered on the 2D plot and each cluster is contained in a separate confidence ellipse then the texture measures used ($\underline{u}$) is regarded as a useful feature for discrimination. The 95 % probability ellipsoid was drawn for each group.

Since $\underline{v}$ is bivariate normal, therefore a value for c may be selected from standard Chi squared tables. The quadratic form, for example, $(\underline{v} - \bar{\underline{v}}_1)^T ((n-1)S_1^{-1})(\underline{v} - \bar{\underline{v}}_1)$ is approximately a Chi squared random variable where the approximation is considered good if $n \geq 25$, (Andrews 1972).

4.2.6 Discrimination Strategy

The main strategy of the current statistical discrimination procedure is to detect normal healthy individuals (NL) first, those not categorized as NL would then be tested as being either PNEU, PTB, or LC.

4.2.6.1 Discrimination Between Disease Present and Disease Absent (NL)

Using two-dimensional wavelet transformation, a selected ROI will generate four components, namely the approximate (LL), vertical (LH), horizontal (HL), and diagonal (HH) components. In each of these four components, twelve texture measures are calculated, yielding 48 descriptors or features. Preliminary explanatory data analysis (EDA) using two-dimensional plots for a given pair of features does not provide obvious separation between clusters. This suggests that the choice of features is non-trivial. Henceforth, a modified principal component (modPC) method is proposed.

In the modPC method, a feature vector, say $\underline{v}$, is selected in two ways. Firstly, each patient's ROI is represented by a four-dimensional vector, for example, entropy LL, entropy LH, entropy HL, and entropy HH. This approach is to investigate the capability of each texture measure as the texture descriptor for each type of disease. The second approach is to use all 12 texture measures (48 dimensional feature vectors). By combining all twelve texture measures, a superior texture descriptor is hoped to be able to represent each disease. In either approach, the appropriate Q matrix is first derived before the principal component (PC) is applied onto it. This method named as the modified PC (modPC) method was successfully applied in the discrimination between PNEU and NL (Noor et al. 2010).

The discrimination strategy between DP and DA, for the three cases PNEU–NL, PTB–NL, and LC–NL required that the feature vector $\underline{v}$ is subjected to the following constrains. In particular, if the DF is such that,

$$DF_{21}(\underline{v}) = \ln \frac{g_2(\underline{v})}{g_1(\underline{v})} > 0 \qquad (4.5)$$

and,

$$DF_{23}(\underline{v}) = \ln\frac{g_2(\underline{v})}{g_3(\underline{v})} > 0 \tag{4.6}$$

and,

$$DF_{24}(\underline{v}) = \ln\frac{g_2(\underline{v})}{g_4(\underline{v})} > 0 \tag{4.7}$$

where the integers 1, 2, 3, and 4 represent PNEU, NL, PTB, and LC, respectively, then $\underline{v}$ is said to represent a patient that has a normal lung.

Note that DF($\underline{v}$) is either the LDF($\underline{v}$) or QDF($\underline{v}$), a choice made after testing for $\Sigma_1 = \Sigma_2$. To reduce the cost of misclassification, three ROIs for each patient were considered. Another purpose of taking three areas for each patient is to ensure that the texture descriptor is homogeneous in the ROI.

4.2.6.2 Discriminating Between Diseases: PNEU, PTB, and LC

Despite the optimistic results of the previous section, any attempt to discriminate PNEU–PTB, PNEU–LC, or PTB–LC will not necessarily yield similar results. This is due to the fact that the abnormal appearance for these diseases on the X-ray film is very similar (Adam and Dixon 2008).

The results showing the lowest misclassification rates are summarized in Table 4.6. Following De Veaux et al. (2009), for a particular pair-wise comparison, the error measure with lowest value will be used as an indicator of performance. In Table 4.6, type II error was used for the PNEU–PTB and PTB–LC comparison, while for the PNEU–LC, type I error was used. The texture measure that gives the lowest error rates for both types of Q matrix will be selected. In particular, when both error rates are similar or equal the choice of Q matrix is arbitrary.

The proposed discrimination procedure is as follows;

Given $\underline{v}$ calculate;

$$DF_{13}(\underline{v}) = \ln\frac{f_{PNEU}(\underline{v})}{f_{PTB}(\underline{v})}, \text{ utilizing } Q_{PTB} \tag{4.8}$$

$$DF_{14}(\underline{v}) = \ln\frac{f_{PNEU}(\underline{v})}{f_{LC}(\underline{v})}, \text{ utilizing } Q_{LC} \tag{4.9}$$

$$DF_{34}(\underline{v}) = \ln\frac{f_{PTB}(\underline{v})}{f_{LC}(\underline{v})}, \text{ utilizing } Q_{PTB} \tag{4.10}$$

Suppose $\underline{v}_o$ is an unknown observation (see Fig. 4.4),

$$\text{If } DF_{13}(\underline{v}_o) > 0 \quad \text{and/or} \quad DF_{14}(\underline{v}_o) > 0 \Rightarrow v_o \in \text{PNEU} \tag{4.11}$$

$$\text{If } DF_{14}(\underline{v}_o) < 0 \quad \text{and/or} \quad DF_{34}(\underline{v}_o) < 0 \Rightarrow \underline{v}_o \in \text{LC} \tag{4.12}$$

$$\text{If } DF_{13}(\underline{v}_o) < 0 \quad \text{and/or} \quad DF_{34}(\underline{v}_o) > 0 \Rightarrow \underline{v}_o \in \text{PTB} \tag{4.13}$$

The above discrimination procedure can be used in two ways. Firstly, the individual Eqs. 4.8–4.10 yields the discrimination problem when only two types of diseases are of interest, for example, Eq. 4.10 should be applied when the interest is only in comparing PTB and LC. The set of conditions determined by Eqs. 4.11–4.13 gives the three population discrimination problem which constitutes the second discrimination procedure.

To increase the test sample size, each ROI is now treated as coming from a separate individual, giving a maximum of 60 counts of possible successful detection (sample size is 20). The selection of ROI was done in similar fashion as in Sect. 4.2.6.1.

4.3 Result

The pair-wise discrimination strategy between disease present (DP) and disease absent (DA) shows that the type I error and type II error were less than 15 % when maximum column sum energy texture descriptor are applied for PNEU–NL, PTB–NL, and LC–NL regardless which Q is used (see Table 4.3). It is interesting to note that when maximum column sum energy is jointly used with the other 11 texture measures the error probability may increase (see last row of Table 4.3). The discrimination procedure using either (or all) of $DF_{21}(\underline{v})$, $DF_{23}(\underline{v})$, or $DF_{24}(\underline{v})$, shown in Fig. 4.3, shows the highest rates of correct classification when utilizing maximum column sum energy and Q_{NL} and is illustrated in Tables 4.4 and 4.5.

For the PNEU–PTB discrimination problem (Table 4.6) using Q_{PTB} and mean of energy and maximum value texture measures give lowest type II error rates. Both texture measures also give lowest error rates for the PNEU–LC and PTB–LC comparisons.

For the three population discrimination problems (Fig. 4.4) only the following case will be discussed;

Select $\underline{v}_o$ from the test set (all three diseases),if

$$DF_{13}(\underline{v}_o) > 0 \quad \text{and} \quad DF_{14}(\underline{v}_o) > 0 \quad \text{implies } \underline{v}_o \in \text{PNEU}$$

else if,

$$DF_{14}(\underline{v}_o) < 0 \quad \text{and} \quad DF_{34}(\underline{v}_o) < 0 \quad \text{implies } \underline{v}_o \in \text{LC}$$

else,

$$DF_{13}(\underline{v}_o) < 0 \quad \text{and} \quad DF_{34}(\underline{v}_o) > 0 \quad \text{implies } \underline{v}_o \in \text{PTB}$$

Table 4.3 Results of pair-wise discrimination for LC–NL, PTB–NL, and PNEU–NL

	Test group (from LDF/QDF)				Test group (from LDF/QDF)				Test group (from LDF/QDF)			
	Qnl		Qlc		Qnl		Qptb		Qnl		Qpneu	
	P(LC|NL)	P(NL|LC)	P(LC|NL)	P(NL|LC)	P(PTB|NL)	P(NL|PTB)	P(PTB|NL)	P(NL|PTB)	P(PNEU|NL)	P(NL|PNEU)	P(Pneu|NL)	P(NL|PNEU)
Mean of energy	0.00	0.20	0.35	0.05	0.90	0.10	0.90	0.10	0.5000	0.2500	0.8000	0.1000
Entropy	0.00	0.15	0.00	0.25	0.60	0.00	0.55	0.30	0.3500	0.3000	0.3000	0.4000
Standard deviation of intensity value	0.00	0.05	0.00	0.15	0.70	0.75	0.70	0.70	0.0500	0.3000	0.0500	0.3500
Standard deviation of energy	0.00	0.20	0.20	0.10	1.00	0.10	1.00	0.10	1.0000	0.0500	0.9000	0.0500
Maximum value	0.00	0.15	0.00	0.10	0.05	0.10	0.00	0.10	0.1000	0.1500	0.1000	0.0000
Minimum value	0.35	0.10	0.30	0.10	0.85	0.05	0.70	0.10	0.7500	0.2000	0.6500	0.2500
Maximum energy	0.05	0.10	0.00	0.10	0.15	0.10	0.20	0.10	0.1500	0.1000	0.1000	0.1000

(continued)

Table 4.3 (continued)

	Test group (from LDF/QDF)				Test group (from LDF/QDF)				Test group (from LDF/QDF)			
	Qnl		Qlc		Qnl		Qptb		Qnl		Qpneu	
	P(LC|NL)	P(NL|LC)	P(LC|NL)	P(NL|LC)	P(PTB|NL)	P(NL|PTB)	P(PTB|NL)	P(NL|PTB)	P(PNEU|NL)	P(NL|PNEU)	P(Pneu|NL)	P(NL|PNEU)
Maximum row sum energy	0.70	0.00	0.70	0.00	0.05	0.15	0.05	0.15	0.1000	0.1500	0.1000	0.1500
Maximum column sum energy	0.05	0.05	0.05	0.05	0.15	0.05	0.15	0.05	0.1500	0.1000	0.1000	0.1000
Zero-crossing	0.00	0.65	0.15	0.90	0.10	0.85	0.30	0.85	0.4500	0.6000	0.0500	0.7500
Contrast	0.90	0.00	0.05	0.10	0.15	0.10	0.25	0.10	0.8500	0.1000	0.1500	0.1500
Homogeneity	0.90	0.00	0.90	0.00	0.95	0.00	0.95	0.00	0.9000	0.0000	0.9000	0.0000
12 Features	0.00	0.10	0.00	0.10	0.30	0.10	0.35	0.10	0.2000	0.1000	0.2500	0.0000

Table 4.4 Correct classification rates for discriminating disease absent (DA) using NL test cases

Texture measure	Maximum column sum energy and Q_{NL}			
Discrimination procedure using NL test data set	$DF_{21} > 0$	$DF_{23} > 0$	$DF_{24} > 0$	$DF_{21} > 0$ and $DF_{23} > 0$ and $DF_{24} > 0$
Percentage correct classification (%)	65	70	98.33	65

Table 4.5 Correct classification results for discriminating disease present (DP) using PNEU, PTB, and LC test cases

Texture measure	Maximum column sum energy and Q_{NL}			
Discrimination procedure	$DF_{21} < 0$	$DF_{23} < 0$	$DF_{24} < 0$	$DF_{21} > 0$ and $DF_{23} > 0$ and $DF_{24} > 0$
PNEU test data set (%)	100	–	–	95
PTB test data set (%)	–	98.3	–	95
LC test data set (%)	–	–	98.3	98.3

When all three conditions are not satisfied the next member of the test set was considered. The above procedure yields 30 % correct classification for PNEU, and 95 % correct classification for PTB and 50 % correct classification for LC.

If the condition in the above procedure is relaxed, for example, if either $DF_{13}(\underline{v}_o) > 0$ or $DF_{14}(\underline{v}_o) > 0$ is satisfied, PNEU was found to be detected with higher correct classification rates (see Tables 4.7, 4.8 and 4.9). In particular, the relaxed condition yields correct classification rates of 70 % for PNEU, 97 % for PTB, and 79 % for LC.

4.4 Discussion

In the medical literature disease detection using chest radiograph is generally confined to the case of comparing (detection) a particular disease with normals. This study investigate the same problem but for three diseases (simultaneously). The results from this study suggest that the proposed statistical discrimination procedure can be used to detect either of PNEU, PTB, and LC when the comparison is made with normals yielding results that are comparable with similar studies (Oliveira et al. 2007; Katsuragawa and Doi 2007; Arzhaeva et al. 2009; Homma et al. 2009).

In the situation where two diseases are suspected, mean of energy or maximum value texture measures can be used for discrimination. Further, the choice of Q matrix is arbitrary.

Table 4.6 Misclassification using LDF/QDF for pair-wise discrimination for PNEU–PTB, PNEU–LC, and PTB–LC

Pair-wise	PNEU–PTB		PNEU–LC		PTB–LC	
Type of Q	Q_{PNEU}	Q_{PTB}	Q_{PNEU}	Q_{LC}	Q_{PTB}	Q_{LC}
Type of error/texture measures	P(type II error) = P (PTB|PNEU)	P(type II error) = P (PTB|PNEU)	P(type I error) = P (PNEU|LC)	P(type I error) = P (PNEU | LC)	P(type II error) = P(PTB|LC)	P(type II error) = P(PTB|LC)
Mean of energy	0.2000	0.2000	0.25	0.25	0.20	0.25
Entropy	0.3000	0.3000	0.25	0.25	0.20	0.20
Standard deviation of intensity value	0.4000	0.2500	0.20	0.35	0.25	0.10
Standard deviation of energy	0.3500	0.4500	0.30	0.35	0.25	0.35
Maximum value	0.6000	0.1500	0.15	0.15	0.15	0.25
Minimum value	0.4500	0.4500	0.25	0.25	0.25	0.25
Maximum energy	0.3500	0.2500	0.25	0.25	0.25	0.25
Maximum row sum energy	0.2000	0.2000	0.25	0.25	0.20	0.20
Maximum column sum energy	0.3000	0.3000	0.30	0.30	0.35	0.35
Zero-crossing	0.4500	0.2000	0.45	0.95	0.20	0.25
Contrast	0.5500	0.4000	0.25	0.20	0.20	0.25
Homogeneity	0.6000	0.4000	0.20	0.25	0.15	0.10
12 Features	0.4500	0.5000	0.25	0.25	0.35	0.15

Table 4.7 Correct classification results for PNEU

Wavelet texture measure	Mean of energy		Maximum value	
Discrimination procedure	$DF_4(\underline{x})$	$DF_5(\underline{x})$	$DF_4(\underline{x})$	$DF_5(\underline{x})$
Correct classification	14/60 (23.3 %)	23/60 (38.3 %)	38/60 (63.3 %)	23/60 (38.3 %)
Combine (satisfy either $DF_4(\underline{x}) > 0$ or $DF_5(\underline{x}) > 0$)	26/60 (43.3 %)		40/60 (66.67 %)	
Combine all (satisfy either $DF_4(\underline{x}) > 0$ or $DF_5(\underline{x}) > 0$ for both texture measures)	42/60 (70 %)			

Table 4.8 Correct classification results for PTB

Wavelet texture measure	Mean of energy		Maximum value	
Discrimination procedure	$DF_4(\underline{x})$	$DF_6(\underline{x})$	$DF_4(\underline{x})$	$DF_6(\underline{x})$
Correct classification	51/60 (85 %)	55/60 (91.67 %)	13/60 (21.67 %)	17/60 (28.33 %)
Combine (satisfy either $DF_4(\underline{x}) < 0$ or $DF_6(\underline{x}) > 0$)	55/60 (91.67 %)		20/60 (33.33 %)	
Combine all (satisfy either $DF_4(\underline{x}) < 0$ or $DF_6(\underline{x}) > 0$ for both texture measures)	58/60 (96.67 %)			

Table 4.9 Correct classification results for LC

Wavelet texture measure	Mean of energy		Maximum value	
Discrimination procedure	$DF_5(\underline{x})$	$DF_6(\underline{x})$	$DF_5(\underline{x})$	$DF_6(\underline{x})$
Correct classification	27/42 (64.3 %)	3/42 (7.14 %)	24/42 (57.1 %)	27/42 (64.3 %)
Combine (satisfy either $DF_5(\underline{x}) < 0$ or $DF_6(\underline{x}) < 0$)	29/42 (69 %)		31/42 (73.8 %)	
Combine all (satisfy either $DF_5(\underline{x}) < 0$ or $DF_6(\underline{x}) < 0$ for both texture measures)	33/42 (78.6 %)			

In the three population discrimination problems, it is not expected for the strict discrimination procedure to yield a high success rate for all three diseases. This must be due to the fact that the diseases may appear to be similar on the chest radiographs. In particular, the procedure must be modified, for example, use other texture measure to allow better detection rates for PNEU and LC. However, if more relaxed constraints on the discrimination procedure were allowed, higher correct classification rates can be obtained. Although the sample sizes for the control data set is 30 and, test data set is 20, the feature vector $\underline{v}$ for all samples (groups) were found to be normally distributed ensuring optimal results when using discriminant function. The results of this research is very promising, however, further work are needed for verification and validation study with larger sample size.

4.5 Conclusion

The proposed novel texture-based statistical discrimination procedure was shown to be able to detect PNEU, PTB, and LC using chest radiograph. The statistical discrimination procedure studied used wavelet texture measure and the modified principal component (modPC) method. The discrimination procedures consist of (1) pair-wise discrimination for disease present (DP) and disease absent (DA), namely PNEU–NL, PTB–NL, and LC–NL, (2) pair-wise discrimination of diseases, namely PNEU–PTB, PNEU–LC, and PTB–LC, and finally (3) a three population discrimination problems for PNEU, PTB, and LC. Low misclassification probability was achieved when maximum column sum energy texture measure is used for (1), and mean of energy and maximum value texture measures were used for (2) and (3). The maximum column sum energy texture measure yielded 98 % correct classification rates for all three diseases. The diseases were then compared pair-wise and the combination of mean of energy and maximum value texture measures gave correct classification rates of 70, 97, and 79 % for pneumonia, PTB and lung cancer, respectively. The results of this research is very promising, however, further work are needed for verification and validation study with larger sample size.

Acknowledgments We would like to acknowledge the contribution from the Director and staff of The Institute of Respiratory Medicine, Malaysia, and Dr Hamidah Shaban, Selangor Medical Centre. This research was funded under an e-Science Fund grant from the Ministry of Science, Technology and Innovation, Malaysia, Universiti Teknologi Malaysia and University of Malaya.

References

Adam A, Dixon AK (eds) (2008) Grainger & Allison's diagnostic radiology a textbook of medical imaging, vol 1, 5th edn. Elsevier, China
Andrews DF (1972) Plots of high dimensional data. Biometrics 28(125):36
Arzhaeva Y, Tax DMJ, Ginneken B (2009) Dissimilarity-based classification in the absence of local ground truth: application to the diagnostic interpretation of chest radiographs. Pattern Recogn 42:1768–1776. doi:10.1016/j.patcog.2009.01.016
Cherian T, Mulholland EK, Carlin JB, Ostensen H, Amin R, de Campo M, Greenberg D, Lagos R, Lucero M, Madhi SA, O'Brien KL, Obaro S, Steinhoff MC (2005) Standardized

interpretation of pediatric chest radiographs for the diagnosis of pneumonia in epidemiological studies. Bull World Health Organ 83(5):353–359

Daubechies I (1992) Ten lectures on wavelets. SIAM, Pennsylvania

De Veaux RD, Velleman PF, Bock DE (2009) Introduction statistics, 3rd edn. Pearson International, Boston, pp 531–584

Frieden T (2004) Toman's Tuberculosis case detection, treatment and monitoring: question and answer. WHO, Geneva

Ginneken B, Katsuragawa S, ter Haar Romeny BM, Doi K, Viergever MA (2002) Automatic detection of abnormalities in chest radiographs using local texture analysis. IEEE Trans Med Imaging 21(2):139–149

Ginnekien B, ter Haar Romeny BM, Viergever MA (2001) Computer aided diagnosis in chest radiography: a survey. IEEE Trans Med Imaging 20(12):1228–1241

Gonzalez RC, Woods RE (1992) Digital image processing. Addison-Wesley, Reading, p 510

Gopi A, Madhavan SM, Sharma SK, Sahn SA (2007) Diagnosis and treatment of tuberculous pleural effusion in 2006. Chest 131:880–889. doi:10.1378/chest.06-2063

Hamilton W, Peters TJ, Round A, Sharp D (2005) What are the clinical features of lung cancer before the diagnosis is made? A population based case-control study. Thorax 60:1059–1065. doi:10.1136/thx.2005.045880

Hara T, Fujita H, Doi K (2007) Computer aided diagnosis in medical imaging: historical, review, current status and future potential. Comput Med Imaging Graph 31(4–5):198–211

Health Facts (2009) Health Informatics Centre, Planning and Development Division, Ministry of Health Malaysia, May 2009

Homma N, Kawai Y, Shimoyama S, Ishibashi T, Yoshizawa M (2009) A study on the effect of morphological filters on computer-aided medical image diagnosis. Artif Life Robot 14:191–194. doi:10.1007/s10015-009-0651-8

Johnson RA, Wichern DW (2007) Applied multivariate statistical analysis, 6th edn. Pearson International Edition, New Jersey

Katsuragawa S (2007) Doi K Computer-aided diagnosis in chest radiography. Comput Med Imaging Graph 31:212–223. doi:10.1016/j.compmedimag.2007.02.003

Malaysian Cancer Statistics (2006) Data and figure peninsular Malaysia 2006, Ministry of Health Malaysia

Mardia KV, Kent JT, Bibby JM (1979) Multivariate analysis. Academic Press, London

Middlemiss H (1982) Radiology of the future in developing countries. Br J Radiol 55:698–699

Moores BM (1987) Digital X-ray imaging. IEE Proc 134(2):115 (Special issues on medical imaging)

Nakamura K et al (1970) Studies on the diagnostic value of 70 mm radiophotograms by mirror camera and the reading ability of physicians. Kekkaku 45:121–128

TB a Problem Once Again. New Straits Time article, 29 March 2008

Noor NM, Rijal OM, Yunus A, Abu-Bakar SAR (2010) A discrimnation method for the detection of pneumonia using chest radiograph. Comput Med Imaging Graph 34:160–166. doi:10.1016/j.compmedimag.2009.08.005

Oliveira LLG, e Silva SA, Ribeiro LHV, de Oliveira RM, Coelho CJ, Andrade ALSS (2007) Computer-aided diagnosis in chest radiography for detection of childhood pneumonia. Int. J Med Inform. doi:10.1016/j.ijmedinf.2007.10.010

Schillham AMR, van Ginneken B, Loog M (2006) A computer aided diagnosis system for the detection of lung nodules in chest radiographs with an evaluation on a public database. Med Image Anal 10(2):246–258

Sonka M, Hlavac V, Boyle R (1998) Image processing, analysis, and machine vision. International Thomson Publishing, Pacific Grove, p 652

van Ginneken B, Katsugarawa S, ter Haar Romeny BM, Doi K, Viergever MA (2002) Automatic detection of abnormalities in chest radiographs using local texture analysis. IEEE Trans Med Imaging 21(2):139–149

Walker JS (1999) A primer on wavelets and their scientific applications. Chapman and Hall/CRC Press, Boca Raton

WHO (2003) International agency for research on cancer. In: Steward BW, Kleihues P (eds). World Cancer Report. WHO, Geneva

WHO (2006) The global plan to stop TB 2006-2015: action for life towards a world free of tuberculosis. WHO, Geneva

WHO Report (2004) Global tuberculosis control, surveillance, planning, financing. WHO, Geneva

Wipf JE, Lipsky BA, Hirschmann JV, Boyko EJ, Takasugi J, Peugeot RL, Davis CL (1999) Diagnosing pneumonia by physical examination: relevant or relic? Arch Intern Med 139:1082–1087

Chapter 5
Imaging of Mitochondrial Disorders: A Review

Sang-Bing Ong

Abstract Mitochondria are the main provider of adenosine triphosphate (ATP) and help in maintaining optimum calcium homeostasis while also participating in cell death cascades. Hence, the fate of cells depends on the optimum functioning and positioning of the mitochondria. Perturbations to the normal functioning of these organelles play a central role in a wide range of mitochondrial diseases, which affect multiple organs with varying severity. Due to this heterogeneity, multiple diagnostic modalities including combinations of clinical, biochemical, and structural criteria have been developed. Imaging techniques such as computed tomography (CT), magnetic resonance imaging (MRI), and magnetic resonance spectroscopy (MRS) have been particularly useful in the diagnosis of mitochondrial diseases at the level of the organ as the central nervous system (CNS) is the second most frequently affected organ, while imaging via microscopy is crucial to detect changes in mitochondria at the cellular level. This review provides a detailed overview of the application of imaging modality in the diagnosis of mitochondrial disorders, from the organ to the cellular level.

5.1 Introduction

5.1.1 The Mitochondrion

Mitochondria are ubiquitous in eukaryotic cells and are the site where oxidative phosphorylation (OXPHOS) for production of energy in the form of adenosine triphosphate (ATP) occurs (Saks et al. 2006; Stanley et al. 2005; Zeviani and Di Donato 2004). Mitochondria are known to have a double membrane and are found

S.-B. Ong, Ph.D. CBiol EurProBiol (✉)
Faculty of Biosciences and Medical Engineering, Universiti Teknologi Malaysia,
Level 3, V01, Block A, Satellite Building, Skudai-Johor, Malaysia
e-mail: sang_bing@yahoo.com

K. W. Lai et al., *Advances in Medical Diagnostic Technology*,
Lecture Notes in Bioengineering, DOI: 10.1007/978-981-4585-72-9_5,
© Springer Science+Business Media Singapore 2014

in the cytosol. Compartmentalization of the mitochondria into (1) the outer mitochondrial membrane (OMM), (2) the inter-membrane space (IMS), (3) the inner mitochondrial membrane (IMM) with the folding of cristae, and (4) the matrix enables the mitochondria to perform specialized functions such as OXPHOS, signaling, mitophagy, induction of apoptosis, and programmed cell division (McBride et al. 2006). A total of 3,000 genes have been documented as fundamental to produce a mitochondrion (DiMauro and Schon 2003; Schapira 2006). Thirty-seven of the genes are encoded by mitochondrial DNA (mtDNA) where the remaining ones are encoded in the nucleus (DiMauro and Schon 2003; Schapira 2006). The proteins produced are transported to the mitochondria (Anderson et al. 1981; Chan 2006a; Ingman et al. 2000). Only 3 % of these genes are responsible for production of ATP, while the remaining genes are involved in other physiological processes such as urea detoxification, cholesterol metabolism, and heme synthesis (Anderson et al. 1981; Chan 2006a; Galluzzi et al. 2012; Ingman et al. 2000; McBride et al. 2006).

5.1.2 What is Mitochondrial Disorder?

The presence of mitochondrial disease, a clinically heterogeneous group of disorders, signifies a failure of the mitochondrion, a key organelle responsible for producing energy in every cell of the body except red blood cells. The failure of this organelle appears to mostly impact cells of the brain, heart, liver, skeletal muscles, kidney, and the endocrine and respiratory systems, albeit in varying levels of severity (DiMauro et al. 1990, 1985; Petty et al. 1986). The criteria for diagnosis of mitochondrial disease should include (1) clinically complete respiratory chain (RC) encephalomyopathy and (2) molecular identification of an mtDNA mutation of undisputed pathogenicity (Bernier et al. 2002). Mitochondrial disease can be caused by either inherited or spontaneous mutations in mtDNA or nuclear DNA (nDNA) (DiMauro and Hirano 2005). These mutations can alter the functions of proteins produced by the mitochondria or even mitochondrial proteins themselves. Identical mtDNA mutations may not produce identical diseases due to the complexity of the interplay between genes and proteins in the cells. Conversely, different mutations can lead to the same diseases (Friedman et al. 2010; McFarland and Turnbull 2009). The typical mitochondrial disorder starts with an isolated organ but often evolves into a multisystem disease. Mitochondrial disease exists in many forms with onset occurring during the congenital stage although adult onset is becoming common (McFarland and Turnbull 2009). Symptoms of the disease include muscle weakness and pain, general fatigue, growth defects, loss of motor control and coordination, gastrointestinal disorders, and susceptibility to infection (Mattman et al. 2011; Morava et al. 2006; Schapira 2006). Proper diagnosis may be masked by pre-conceived assumptions that mitochondrial diseases may be other diseases. Most of the mitochondrial diseases in patients are not detected at an early stage due to the complexity of the disease.

5.1.3 Mitochondrial Disease Clinical Manifestations: An Overview

Mitochondrial diseases produce a plethora of manifestations and can be present at any age. The manifestations of mitochondrial diseases range from acute metabolic derangement to intermittent episodes of dysfunction to gradual progressive neurodevelopmental decline or regression. Understanding the clinical manifestations of pediatric and adult-onset mitochondrial disease will help family physicians to properly diagnose and treat the patients (Mattman et al. 2011).

5.1.3.1 General Characteristics of Pediatric and Adult-Onset Disease

The pediatric mitochondrial disease ranges from lethargy, hypotonia, failure to thrive, seizures, cardiomyopathy, deafness, blindness, movement disorder, and lactic acidosis to progressive neurological, cardiac, and liver dysfunction (DiMauro et al. 1985, 1990; DiMauro 2004). The patients suspected of harboring mitochondrial diseases based on the presence of these symptoms should be referred to a tertiary care center for proper evaluation and diagnosis. Manifestations of the symptoms may vary between family members with a maternally derived history of illness (Mattman et al. 2011).

Adult-onset mitochondrial diseases vary in the presentation times and can be triggered off during adulthood or discovered in adulthood following childhood symptoms (DiMauro et al. 1985; Mattman et al. 2011; Zeviani and Di Donato 2004). Adult-onset mitochondrial disease is typically a progressive multisystem disorder and is often discovered following physical examination and laboratory evaluation, even though the initial onset may have happened in a specific organ. The presence of a mitochondrial disease should be suspected when specific clinical manifestations are present and accompanied by one or more of the following: (a) involvement of different organ systems and/or (b) abnormal severity (i.e., early onset with progression over time), and/or (c) maternal inheritance pattern (Mattman et al. 2011).

5.2 Imaging and Diagnosis of Mitochondrial Diseases

Diagnosis of mitochondrial diseases is complicated by the complex interaction between the nuclear genome and mitochondrial genome (Zeviani and Di Donato 2004). Solely relying on enzymological studies is not sufficient to differentiate between mitochondrial and nuclear origins of diseases (Rötig et al. 2004). An alternative would be to study the phenotype of the resulting fused cell between the patient's fibroblast and mtDNA-null (p^0) fibroblast (Rötig et al. 2004). Nevertheless, this approach is not feasible as it can only be performed in patients expressing the RC deficiency in the fibroblasts, in addition to the high-cost and time-consuming protocol. The central nervous system (CNS) is the second most frequently affected organ in mitochondrial disorders, alone or in combination (Finsterer et al. 2001). The abnormalities detected during brain imaging can indicate metabolic disease, albeit non-specific. The varying stages of mitochondrial diseases further add to the complexity of proper diagnosis (Bernier et al. 2002; Morava et al. 2006). A combination of modalities for diagnostic confirmation is required due to the presence of non-specific symptoms and mixed etiologies (Haas et al. 2007; Thorburn et al. 2004). The most commonly used and clinically relevant imaging techniques for visualization of CNS abnormalities in mitochondrial disorders include computed tomography (CT), magnetic resonance imaging (MRI), and magnetic resonance spectroscopy (MRS) (Finsterer 2009).

5.2.1 Computed Tomography (CT) Scans

Computed tomography (CT) was the primary neuroimaging method prior to the advent of MRI and was mainly used for the detection of calcification, a pathophysiological process of mitochondrial disease, focal or diffuse atrophy, ischemic lesions, white matter lesions, and demyelination (Finsterer 2009). Intracranial calcification has been detected in a variety of disorders including idiopathic basal ganglia calcifications and spinocerebellar ataxia 20 (Knight et al. 2004). Extensive calcification on CT scans has also been detected in a patient with a novel polymerase gamma 1 (*POLG1*) mutation (Sidiropoulos et al. 2013). Nevertheless, CT scans face limitations when used to visualize certain abnormalities, such as strokelike lesions, symmetric necrosis of the thalami, basal ganglia, diencephalon, or brain stem in patients with Leigh's syndrome (Finsterer 2009; Iizuka et al. 2002; McFarland et al. 2002). Current diagnostic protocols rely on the detection of changes in the neurological system at the initial stage, followed by confirmation via the usage of MRI in the CNS. The confirmation via MRI relies on the detection of key characteristics feature of the disease and non-specific abnormalities (Barragán-Campos et al. 2005; Muñoz et al. 1999). Detailed measurements of the changes in chemicals in the brain as indices for disorder characterization can be performed using MRS. The usage of MRI and MRS is extremely safe and non-invasive, hence rendering them suitable for monitoring of disease progression and changes in metabolic markers.

5.2.2 Magnetic Resonance Imaging (MRI)

Structural abnormalities in mitochondrial disorders are detected by MRI. The most common manifestation of mitochondrial disease is a global delay in myelination pattern at the initial stage of the disease—an indication of abnormal metabolic process, which is only resolved at a later stage following development (Dinopoulos et al. 2005; Muñoz et al. 1999). Hypointensity on T1 images as well as a symmetric signal abnormality of deep gray matter presenting with hyperintensity on T2 and fluid attenuation inversion recovery (FLAIR) images constitute the most common specific MRI findings indicative of mitochondrial disease (Gire et al. 2002; Haas and Dietrich 2004). Strokelike lesions affecting the white matter and gray matter are regarded as manifestations of a vasogenic edema and show dynamic changes in intensity and extensiveness over varying periods of time (Iizuka et al. 2007). Diffusion-weighted imaging (DWI) is considered to be more sensitive than T2-weighted images to demonstrate strokelike lesions (Abe 2004). The lesions can be either patchy or homogeneous with the presence of varying degrees of cerebral and cerebellar atrophy (Saneto et al. 2008). Cerebral atrophy may be categorized as focal or diffuse, cortical or subcortical, supratentorial or infratentorial, and primary or secondary after strokelike lesions (Finsterer 2009). Some patients may develop cerebellar atrophy with or without cerebral atrophy (Scaglia et al. 2005; Van Goethem et al. 2004). These MRI findings are mostly associated with syndromic phenotypes as discussed in Sect. 2.4. MRI, by itself, however, still lacks adequate sensitivity or specificity for an accurate diagnosis of certain mitochondrial diseases and is more commonly paired with other imaging modalities such as MRS for determining mitochondrial diseases (Barkovich et al. 1993; Friedman et al. 2010; Valanne et al. 1998).

5.2.3 *Magnetic Resonance Spectroscopy*

Magnetic resonance spectroscopy (MRS), capable of measuring biochemical changes in the form of metabolites possessing resonating nuclei (hydrogen-1, ^{1}H; phosphorous-31, ^{31}P; carbon-13, ^{13}C) in the mM range, provides valuable metabolic information as a complement to conventional MRI (Saneto et al. 2008). The ^{1}H-MRS is advantageous in that it shares a similar radiofrequency range compared with conventional MRI and thus can be performed simultaneously in the same examination (Saneto et al. 2008). ^{1}H-MRS aids in predicting and classifying childhood white matter diseases, brain creatine deficiency syndromes, and phenotypic mitochondrial diseases (Chi et al. 2011a). The other nuclei, such as phosphorous-31 (^{31}P) and carbon-13 (^{13}C), although used to measure phosphocreatine and metabolite production degradation rates, respectively, are less utilized due to (1) the need of specific hardware for measurement, (2) lower signal to noise per unit time, and (3) the requirement of costly labeled substrate (^{13}C glucose or acetate) as in the case of ^{13}C MRS (Saneto et al. 2008).

Approximately 30 or more brain metabolites can be distinguished using specialized ^{1}H acquisition approaches compared with only about 5–10 brain metabolites using routine clinical ^{1}H-MRS (Ross 2000). Using specific spatial acquisition techniques such as point-resolved spectroscopy (PRESS), stimulated echo acquisition mode (STEAM), or image selected in vivo spectroscopy (ISIS)

(Keevil 2006), grid (multivoxel also referred to as chemical shift imaging: [CSI or MRSI]) acquisitions can be obtained, although longer acquisition times will be needed to obtain sufficient signal-to-noise ratio (Saneto et al. 2008).

MRS signals are shown as spectra with different peaks corresponding to the various specific chemical entities. The magnetic field strength produced by nearby electrons for the nucleus determines the position of each resonance peak in the spectrum. The signal shifts to a lower frequency when a higher density of electrons is present. Conversely, a signal of higher frequency is present when the density of electrons is lowered (Saneto et al. 2008). J-coupling refers to the interaction between electrons on close neighboring protons within a molecule. Depending on the acquisition delay (echo times or TE) which can be modified to 20–30, 135–144, and 288 ms, the peak of a particular metabolite can be altered by j-coupling, which is advantageous to detection in clinical ^{1}H-MRS (Saneto et al. 2008). As an example, the interaction between two protons (OH–H) with the magnetic field creates a characteristic doublet for the lactate peak at 1.33 parts per million (ppm) (Rand et al. 1999; Saneto et al. 2008). Depending on the echo time, the doublet can be either pointing up (TE = 288 ms) or pointing down (TE = 144 ms) and this specific feature can be utilized for unequivocal measurement of lactate, compared with the overlapping peaks from lipid/macromolecule signal in brain (Rand et al. 1999; Saneto et al. 2008; Yablonskiy et al. 1998). Short TE (20–30 ms) is more useful for the detection of glutamine, glutamate, and myo-inositol although the baseline value may include lipid signals between 1 and 2 ppm (Yablonskiy et al. 1998). Using a higher TE, the lipid peaks can be ruled out while providing a lower signal-to-noise ratio (S/N) (Yablonskiy et al. 1998). The relationship between TE and magnetic field strength, tesla (T), is also crucial for sensitivity of MRS. An improved sensitivity for the spectra at 3.0 T was detected compared with 1.5 T at short TE (Barker et al. 2001). Conversely, the 3.0 T MRS has a reduced or absent signal intensity using an intermediate echo time (Lange et al. 2006). No differences were noted using a long TE, irrespective of magnetic field strength (Barker et al. 2001).

Single-voxel (one region, typically 2–10 cc's) technique can be advantageous for specific localization in areas near the air/brain interfaces (e.g., cerebellum) and optimal signal to noise per unit time, but it has the downside of limited spatial coverage (Saneto et al. 2008) as well as the possibility of underestimating the lactate concentration (José da Rocha et al. 2008). Multiple voxel acquisition (in a single plane) is more commonly performed using chemical shift imaging techniques. The grids of data in matrices of (1) 16 × 16, (2) 24 × 24, or (3) 32 × 32 having individual voxels of approximately 1–2 cc after reconstruction filtering confer the benefit of being able to study many different brain regions at the same time, particularly in regions with inconsistent patterns of lactate elevations (Saneto et al. 2008). The downside is that individual voxels have to be summed to facilitate measurement of lactate with sufficient signal-to-noise ratio. The measurement in voxels may be further complicated with the difference in visibility of lactate in different compartments, particularly when only a partial volume is sampled (Saneto et al. 2008).

5.2.3.1 MRS in Mitochondrial Diseases: Lactate and NAA Changes

MRS is used to diagnose mitochondrial disease based on the changes in different markers following impaired OXPHOS or disruption of electron transport chain (ETC). The shift of catabolic metabolism from the TCA cycle to anaerobic glycolysis produces cerebral lactate in the range of 3–11 mM (Wilichowski et al. 1999) or an average of ∼6 mM (Isobe et al. 2007; Saitoh et al. 1998), which can be measured along with other markers of cellular integrity and energetic such as myo-inositol, choline, creatine, and N-acetylaspartate (NAA). Cellular compromise is almost always denoted by increased lactate and decreased NAA (Jeppesen et al. 2003; Moroni et al. 2002). Due to disease variability and regional sampling, lactate elevations may not be detected in some of the patients, including patients with a normal brain MRI (Bianchi et al. 2003; Chi et al. 2011b; Lin et al. 2003). Cady et al. showcased the possibility of confusion in identification of lactate with propan-1,2-diol, localized at 1.1 ppm (Cady et al. 1994). Most of the increased lactates are detected in lesions, and the intensities of the lactate peaks change according to severity of the disease and the developmental course of lesions (Bianchi et al. 2003; Lin et al. 2003). Saneto et al. (2008) recommend using multivoxel acquisition for MRS, followed by reformatting to identify regions of interest so as to study multiple brain sites and compare contralateral regions. Monovoxel acquisition is useful to detect the lactate spikes in the putamen and the cerebellar dentate nucleus (Delonlay et al. 2013). As opposed to the elevation of lactate levels, a reduction in lactate levels can also be detected depending on the type of mitochondrial disease, time course of disease development, or location of the brain sampled (Saneto et al. 2008).

The presence of elevated lactate is almost universally indicative of neuronal/axonal compromise and decreases in NAA. In a survey of mitochondrial patients, 93 % showed NAA/creatine ratio reductions in cerebellum, 87 % in cortical gray matter regions, though few changes were observed in white matter (Bianchi et al. 2003). Dinopoulos et al. (2005) showed that 11 patients with definitive mitochondrial disease had decreased NAA/creatine ratio in both gray matter and white matter.

Voxel placement is crucial for the detection of abnormal lactate and can be missed if the voxel is placed over the incorrect brain region. This is evident in the case study of Boddaert et al. (2008) where lactate peaks were detected in 9 out of 11 patients with cerebellar involvement when voxels were placed over the cerebellum, while only 3 out of 11 showed lactate voxels over the putamen. In addition, elevated levels of alanine, glucose, or pyruvate can also be detected due to monoxidative glycolysis or pyruvate dehydrogenase complex deficiency (Medina et al. 1990; Wilichowski et al. 1999).

To summarize, although MRS can complement conventional MRI in studying the underlying biochemical changes, the optimal protocol for determining the abnormalities in the brain has yet to be clearly defined.

5.2.4 Mitochondrial Disorders

5.2.4.1 Leigh's Syndrome

Leigh's syndrome or subacute necrotizing encephalomyelopathy is a progressive degenerative disorder caused by mtDNA or nDNA mutations that affect infants, children, and sometimes adults (DiMauro 2004). The most common biochemical abnormality of Leigh's syndrome lies in the defective complex IV (cytochrome C oxidase) (Rahman et al. 1996). Clinical presentation includes global developmental delay, feeding and swallowing difficulties, vomiting, spasticity, brainstem dysfunction, dystonia, abnormal eye movements, and multiple organ involvement (Barkovich et al. 1993; Rahman et al. 1996).

The disease is named after Denis Leigh who first described the neuropathological features of the disease, which include focal, bilateral, and symmetric necrotic lesions associated with demyelination, vascular proliferation, and gliosis in the brain stem, diencephalon, basal ganglia, and cerebellum (Leigh 1951). Progressive signal abnormalities with the highest frequency in the lentiform nuclei and caudate nuclei are detected with MRI. Besides that, abnormalities involving the thalamus, periaqueductal gray, tegmentum, red nuclei, and dentate nuclei can also be commonly seen (Saneto et al. 2008). The spongiform changes and vacuolation in the deep gray structures are reflected by the high T2 signal of MRI. Extensive gliosis and cystic degeneration may occur in the white matter. Compared to the cerebral organic acidurias that lead to severe global degeneration, the volumes of the lesion(s) are actually preserved on follow-up studies. In some cases of Leigh's syndrome, a marked global atrophy over the course of time can be detected even though lesions in the basal ganglia are preserved. As the disease progresses, the basal ganglia can also show volume loss (Saneto et al. 2008).

The etiologies of Leigh's syndrome can be differentiated by using MRI. MRI scans of patients with SURF-1 (a major gene associated with complex IV deficiency) mutations reveal symmetric lesions in the brain stem, subthalamic nuclei, and possibly cerebellum, while some of the patients show basal ganglia abnormalities as well as signal hyperintensities in bilateral optic radiation on T2-weighted imaging and DWI (Farina et al. 2002; Tanigawa et al. 2012; Zhu et al. 1998). Conversely, patients with Leigh's syndrome from other etiologies exhibit T2 hyperintensities in the putamina with involvement of the caudate nuclei, globus pallidi, thalami, and brain stem (Farina et al. 2002; Quinonez et al. 2013; Valanne et al. 1998). Furthermore, there were less symmetric areas on increased T2 signal in the cortex and the subcortical white matter of the right and left occipital lobes and in the perirolandic region (Quinonez et al. 2013).

A lactate doublet peak on MRS and hyperperfusion in the patients at baseline with persistent hyperintensities on DWI at follow-ups were also reported (Chen et al. 2012b). In the absence of hypoxia, ischemia, or infection, the presence of symmetric deep gray structures or lactate peaks in the area of MRI abnormality on 1H-MRS warrants further investigation for mitochondrial defects. High choline levels were also detected in the white matter by MRS (Sijens et al. 2008). A diffused supratentorial leukodystrophy involving the deep lobar white matter can also be detected by the MRI in a small cohort of patients with Leigh's syndrome. Initial swelling followed by cystic degeneration from the posterior to anterior has been reported (Lerman-Sagie et al. 2005; Moroni et al. 2002).

5.2.4.2 Mitochondrial Myopathy, Encephalopathy, Lactic Acidosis, and Strokelike Episodes (MELAS)

Mitochondrial myopathy, encephalopathy, lactic acidosis, and strokelike episodes (MELAS) are maternally inherited progressive neurodegenerative disorders with various symptoms such as headache with nausea preceding strokelike events, treatment-resistant partial seizures, short stature, muscle weakness, exercise intolerance, deafness, diabetes, and slow progressive dementia (Hirano and Pavlakis 1994; Pavlakis et al. 1984). In addition to different mitochondrial and nDNA mutations (Emmanuele et al. 2013; Lamperti et al. 2012; Rossmanith et al. 2008; Sproule and Kaufmann 2008; Tam et al. 2008), an estimated 80 % of MELAS patients have an adenine-to-guanine transition at the tRNA for leucine at position 3,243 in the mtDNA (Goto et al. 1990; Janssen et al. 2006; Lamperti et al. 2012; Malfatti et al. 2007). This results in impaired ATP production, following failed synthesis of mitochondrial proteins.

Hallmark MRI features for MELAS patients include transient strokelike lesions predominantly affecting gray matter with the occasional diffused white matter lesions involving periventricular white matter, centrum semiovale as well as the corpus callosum (Apostolova et al. 2005; Barkovich et al. 1993; Conway et al. 2011; Hirano and Pavlakis 1994; Matthews et al. 1991). Areas of hypoattenuation in the left occipital and partially parietal lobe (Pauli et al. 2013) as well as bipallidal microcalcifications (Renard et al. 2012) have been detected using CT scans. Large bilateral hypointensities in the pallidum and the substantia nigra were detected using 3-T T2-weighted MRI (Renard et al. 2012). Strokelike lesions (accompanying strokelike symptoms) can be seen on MRI at different stages of the disease (Renard et al. 2012). Similarly, brain-MRS-based detection of lactate, increased alanine and glucose, decreased N-acetylaspartate, choline, and creatine can be used to resolve the different stages of MELAS progression and treatment (Chen et al. 2012a; José da Rocha et al. 2008).

The difference between chronic lesions and acute ischemic stroke episodes lies in the diffusion coefficient. Intense T2 and FLAIR signals are common features of both events. However, chronic lesion displays a normal to slightly increased apparent diffusion coefficient (ADC) due to vasogenic edema (Oppenheim et al. 2000; Pauli et al. 2013; Yonemura et al. 2001), while acute ischemic stroke episode has a significant reduction in ADC due to cytotoxic edema (Pauli et al. 2013). Therefore, the DWI sequences during the acute event are fundamental to differentiating between the chronic nature and acute nature of the lesions.

Mutations in the POLG1 gene encoding for the DNA polymerase γ, pol γ which replicates the human mitochondrial genome, produce MELAS-like symptoms (Cheldi et al. 2013; Deschauer et al. 2007). Nevertheless, it is difficult to directly compare between MELAS- and POLG1-associated encephalopathy, which has a heterogeneous clinical presentation (Cheldi et al. 2013). In addition, there have also been documented cases of sole involvement of the basal ganglia as well as ETC defects with strokelike brain lesions that do not fulfill the criteria for full MELAS (Kim et al. 2001), hence warranting the need for multiple diagnostic modalities beyond imaging alone to ensure proper and correct diagnosis of MELAS. MELAS syndrome has also been documented to mimic the clinical and radiological signs of herpes simplex encephalitis (Gieraerts et al. 2013). Effective differentiation was achieved by observing diffusion restriction in some parts of the lesions but not throughout the entire lesions following DWI and biochemical investigations on cerebrospinal fluid, electromyogram, muscle biopsy, and genetic analysis (Gieraerts et al. 2013).

5.2.4.3 Pearson's Syndrome/Kearns–Sayre's Syndrome

Pearson's syndrome, first described in 1979, is an uncommon, multisystem mitochondrial disorder caused by single mtDNA deletions (Lee et al. 2007; Pearson et al. 1979). Pearson's syndrome was first characterized in 4 patients with refractory sideroblastic anemia and vacuolization of marrow precursors and exocrine pancreatic dysfunction (Pearson et al. 1979). Following this characterization, small deletions of 4,799 and 5,500 base pairs in the mtDNA have been shown to be responsible for Pearson's syndrome (McShane et al. 1991; Rotig et al. 1989).

Clinical manifestations of Pearson's syndrome include hypotonia, developmental delay, ataxia, refractory sideroblastic anemia, pancytopenia, exocrine pancreatic dysfunction, and sometimes hepatic or renal failure (Lee et al. 2007). Premature death during infancy or childhood following infections or metabolic failure often occurs, whereas those that do survive into adulthood develop Kearns–Sayre's syndrome (KSS), even with the initial spontaneous remission of infantile sideroblastic anemia (Larsson et al. 1990; Lee et al. 2007; Rotig et al. 1989).

There is relatively little information on CNS imaging of patients with Pearson's syndrome. In 2007, Lee et al. documented a review study of 55 cases of patients with Pearson's syndrome. Neuroimaging findings for Pearson's syndrome were variable, with a myriad of normal, non-specific finding or prominent abnormal signal intensity over the white matter, basal ganglion, or brain stem (Lee et al. 2007). Cortical atrophy has also been reported (Lee et al. 2007).

Kearns–Sayre syndrome (KSS) is a progressive external ophthalmoplegia present before 20 years of age and is commonly associated with atypical retinal pigmentary degeneration and heart cardiac conduction defects (Kearns and Sayre 1958). Cerebellar ataxia, deafness, diabetes, short stature, hypoparathyroidism, and other endocrinopathies may also occur (Mayer et al. 2011). The presence of heteroplasmic rearrangements of multiple DNA deletions in mtDNA denotes the presence of this syndrome (Zeviani et al. 1988).

Typical histopathological findings of this syndrome include status spongiosus involving both gray and white matters, specifically the brainstem tegmentum, basal ganglia, and white matter of the cerebrum and cerebellum (Chi et al. 2011; McKelvie et al. 1991; Tanji et al. 1999). Calcifications are commonly detected in the basal ganglia (José da Rocha et al. 2008). Commonly detected MRI characteristics include cerebral and cerebellar atrophy with early T2/FLAIR hyperintense bilateral lesions in subcortical white matter, thalamus, basal ganglia (substantia nigra and globus pallidus), and brain stem (Chu et al. 1999; Duning et al. 2009; Hourani et al. 2006; Kamata et al. 1998; Leutner et al. 1994; Wray et al. 1995). Cardiovascular MRI has also revealed a potentially typical pattern of diffuse intramural late-gadolinium-enhancement in the left ventricular inferolateral segments of patients suffering from KSS (Yilmaz et al. 2012). Lactate peak on the right putaminal lesion can be detected by MRS (José da Rocha et al. 2008).

5.2.4.4 Alpers' Syndrome

Alpers' syndrome, first described in 1931, is a severe hepatocerebral disease caused by different types of mtDNA depletion that is present at various ages depending on the type of mutation within the polymerase gamma 1 gene (POLG) (Alpers 1931; Huttenlocher et al. 1976; Naviaux and Nguyen 2004). Compound heterozygote mutations, usually one of which within the linker region, cause a more severe disease phenotype with onset before 2 years of age, whereas patients with homozygous mutations within the linker region usually have a later onset and milder form of Alpers' syndrome (Ashley et al. 2008; Saneto et al. 2008; Uusimaa et al. 2013). Mutations in the mitochondrial replicative helicase Twinkle (Hakonen et al. 2007) or the heterozygous presence of a spacer region mutation in trans with another recessive POLG mutation (Kurt et al. 2010) has also been shown to demonstrate phenotypes reminiscent of Alpers' syndrome, thus suggesting that there may be other etiologies as well. Nevertheless, the current definitive diagnosis for Alpers' syndrome is the identification of mutations in POLG (Hunter et al. 2011).

Patients with Alpers' syndrome usually suffer from refractory seizures, psychomotor regression, and a characteristic liver disease (Naviaux and Nguyen 2004). Depending on the site of the POLG gene mutation, the onset and severity vary from infancy to adulthood, while infantile Alpers' syndrome can slowly progress to external ophthalmoplegia and ataxia (Horvath et al. 2006).

The syndrome is denoted by extensive gliosis and neuronal loss in the occipital cortices and cerebellar cortex (Purkinje cells). In addition, there is necrosis of subcortical deep nuclei, hippocampi, lateral geniculate body of the thalamus, and amygdala (Harding 1990). There are also T2/FLAIR hyperintensities in the MRI scans of patients within the occipital regions, deep cerebellar nuclei, thalamus, and basal ganglia (Tzoulis et al. 2006), albeit these rely on the mutations present.

Compound heterozygotes are more involved, while homozygotes have lesions that can reverse during quiescent periods.

In the study of Saneto et al. (2008), they found that the MRI lesions evolved initially from the cerebellum to the occipital regions and pre-motor cortex in one of the patients with compound heterozygous mutations in polymerase gamma 1 (p.Q67X and p.A467T). The patient exhibits clinical symptoms such as focal seizures, nystagmoid eye movement and visual hallucinations (Saneto et al. 2008). Following a rapid course of decline, the patient became blind and developed liver failure after valproic acid exposure. In the same study, a patient with a homozygous mutation in the linker region of polymerase gamma 1 (p.A467T) developed mild sensorineural hearing loss, ataxia, and clumsiness at 5 years of years (Saneto et al. 2008). This patient then developed partial status epilepticus at the age of 15 years. A stable quiescent state for 6–7 years ensued. The patient passed away at the age of 23 years from liver failure, following valproic acid exposure (Saneto et al. 2008). Although the patient's MRI findings were initially normal, occipital T2/FLAIR hyperintensities were detected at the time of partial status epilepticus which also resolved over time, rendering a virtually normal MRI study just before her death (Saneto et al. 2008). Based on this study, it can be concluded that although MRI may be useful in differentiating the severity of Alpers' syndrome, one should always be wary of the possible changes in MRI scans and further studies are needed to correlate MRI findings with corresponding disease phenotype and genotype.

The review of the case records of 12 patients with Alpers' syndrome by Hunter et al. (2011) showed CT scans of focal areas of low attenuation or diffuses cerebral atrophy in four out of eight children scanned. Ten MRI scans of four children dating back to 1992 showed focal high signal changes on T2-weighted images, albeit the positioning of the abnormalities varies. MRS scans were not performed for those patients (Hunter et al. 2011). According to the findings of McCoy et al. (2011), diffusion abnormalities correlated with the T2 abnormalities were acute rather than chronic, while MRS demonstrated the presence of lactate, suggesting a RC defect. Most recently, Uusimaa et al. (2013) also elaborated the importance of increased lactate on MRS and suggestive brain MRI changes (with thalamic predominance) as the impetus for screening of the common POLG mutations and POLG sequencing for the detection of Alpers' disease.

5.2.4.5 Mitochondrial Neuro-Gastrointestinal Encephalomyopathy

Mitochondrial neuro-gastrointestinal encephalomyopathy (MNGIE), first described in 1983, is derived from mutations within the thymidine phosphorylase gene (TYMP) located on chromosome 22q13.32 and causes loss of mtDNA via DNA deletions and point mutations within mtDNA (Ionasescu 1983; Nishino 1999; Nishino et al. 2000) although novel pathogenic mutations have also been reported (Libernini et al. 2012). Intestinal pseudoobstruction, recurrent diarrhea, nausea, and vomiting are the most dominant symptoms. Other symptoms such as ophthalmoparesis, ptosis, and peripheral neuropathy may also be present. The symptoms generally appear before 20 years of age, and younger patients typically have more severe symptoms with death occurring by 35 years of age for approximately 50 % of patients. There have been cases of patients remaining undiagnosed for many years who later undergo extensive workup to Crohn's disease (Perez-Atayde 2013).

Prominent leukoencephalopathy can be detected via MRI in almost all patients. The changes in white matter are diffused with increased T2 signal abnormalities, with a case of widespread supratentorial cortical atrophy described in one study (Barragán-Campos et al. 2005). Mild white matter edema or mild demyelination has also been shown in postmortem studies (Bardosi et al. 1987). The presence of megamitochondria in gastrointestinal ganglion cells and in smooth muscle cells of muscularis mucosae and muscularis propria also aids in the diagnosis of MNGIE (Perez-Atayde 2013).

There have been reported cases of observance of MNGIE-like phenotypes in patients with mutations in the ribonucleoside diphosphate reductase subunit M2 B gene (RRM2B) (Shaibani et al. 2009) as well as the POLG1 gene (Tang et al. 2011, 2012; Goethem et al. 2003). Brain MRI of the RRM2B patient showed patchy leukoencephalopathy and also bilateral basal ganglia signal abnormality compared with the commonly observed diffuse leukoencephalopathy in MNGIE due to TYMP mutations (Shaibani et al. 2009). Depending on the mutations in the POLG1 gene, findings from brain MRI range from normal with no leukoencephalopathy (Goethem et al. 2003) to global cerebral atrophy or increased T2 signal in the basal ganglia (Tang et al. 2012).

5.2.4.6 Isolated Electron Transport Chain Disorders

The mitochondrial and nuclear genomes encode for subunits of all RC complexes, except complex II (Rötig et al. 2004). Nuclear genetic defects in OXPHOS and non-OXPHOS proteins involved in the assembly or maintenance of the ETC are the main cause of pediatric cases of mitochondrial disease (Salvatore DiMauro 2004; Shoubridge 2001). Diagnosis is hampered by the clinically heterogeneous nature of these disorders, mainly caused by the interplay of mtDNA and nDNA expression.

Non-specific diffused white matter changes have been found in some patients expressing ETC defects, involving individual and combinations of complexes I–IV deficiencies (Lerman-Sagie et al. 2005). Progressive macrocystic leukodystrophy or diffuse white matter disease including cerebellar white matter loss has been described in four patients with defects in complex I (Schuelke et al. 1999). In the study of Lebre et al. (2011) consisting of 30 patients with complex I deficiency, bilateral brainstem lesions were observed in all cases with 23 patients exhibiting anomalies of the putamen. Patients harboring mtDNA mutations showed supratentorial strokelike lesions, whereas necrotizing leukoencephalopathy was observed in patients with nDNA mutations (Lebre et al. 2011). Baertling et al. (2013)

showcased the cerebral MRI with bilateral cystic lesions in the centrum semiovale from an 8-month-old infant girl with rapid developmental regression, who was found to harbor an isolated mitochondrial complex I deficiency due to an NDUFS1 mutation (encoding NADH-dehydrogenase-ubiquinone Fe–S protein 1). Demyelination of the supratentorial white matter ensued in the follow-up imaging after 3 months (Baertling et al. 2013).

Complex II, also known as succinate: ubiquinone oxidoreductase or succinate dehydrogenase, is exclusively nuclear-encoded and participates in the citric acid cycle by oxidizing succinate to fumarate(Rutter et al. 2010; Sun et al. 2005). In the mitochondrial ETC, complex II functions to shuttle electrons to ubiquinone (Rutter et al. 2010; Sun et al. 2005). Various studies also reported neurological disorder with leukoencephalopathy, Leigh's syndrome, or cerebellar atrophy in isolated complex II defect (Brockmann et al. 2002; Burgeois et al. 1992; Jain-Ghai et al. 2013; Moroni et al. 2002). In these studies reported, MRI shows abnormal intensities in various sections of the brain as well as multiple foci of diffusion restriction with symmetric distribution (Brockmann et al. 2002; Burgeois et al. 1992; Jain-Ghai et al. 2013; Moroni et al. 2002).

Among two patients with defects in complex III, one had a mutation in cytochrome b subunit of complex III and the other was without a recognized mutation but a defect in enzyme activity (Majoie et al. 2002). White matter changes have been described in at least 5 patients with defects in complex IV (Jaksch et al. 2001; Rahman et al. 2001). Similarly, rapid progression of cerebral/cerebellum atrophy on axial T2-weighted and FLAIR are well visualized in complex IV disorders (Castro-Gago et al. 1999; Scaglia et al. 2005; Valanne et al. 1998). A patient having multiple ETC defects coupled with leukodystrophy has also been reported (Moroni et al. 2002). Leukodystrophy in patients is usually detected at infancy or at a very young age. Multisystem disorders with a neurodegenerative course can be present as well. MRI detection of a diffuse leukodystrophy within the clinical context of multisystem involvement compels the need for further investigation for a mitochondrial disorder. In terms of MRS, lactate elevation is a consistent feature of patients with various ETC defects (Dinopoulos et al. 2005).

5.3 Imaging at the Cellular Level

The existence of healthy mitochondria in the cell relies on maintenance of proper morphology, membrane potential, and calcium signaling, while dysfunctional or senescent mitochondria are 'pruned' by mitophagy. Any deviation will result in mitochondrial disorder which will ultimately lead to a disease state. Hence, an understanding of the mitochondrial function in normal and pathophysiological states is critical for the monitoring and treatment for mitochondria-related diseases. Methods for imaging the changes in mitochondria at the cellular level have been relatively well established using epifluorescence, confocal laser, and electron microscopy (Chan 2006b; Chen and Chan 2004; Gieraerts et al. 2013; Karbowski et al. 2004; Ong and Gustafsson 2012; Ong and Hausenloy 2010; Ong et al. 2010; Rötig et al. 2004). More recently, multiphoton microscopy has also been explored for imaging of mitochondrial function (Davidson and Duchen 2012; Hall et al. 2009, 2013). In the following sections, I will review some of the latest innovations in mitochondrial imaging at the cellular level.

5.3.1 Imaging Mitochondrial Dynamics and Its Disorders

Research for the past decade has revealed a unique feature of the mitochondria, in that the mitochondria exist in a dynamic equilibrium between a fused (elongated) (Figs. 5.1a, 5.2) and fragmented (Fig. 5.1b) state (Chan 2006b; Chen and Chan 2004; Liesa et al. 2009; Ong et al. 2012; Ong and Hausenloy 2010).

The movement of the mitochondria and subsequent changes in shapes (morphology) of the mitochondria constitute the term of 'mitochondrial dynamics' and have since led to an impetus of research in this particular area. The number and shapes of mitochondria vary according to cell type and energy requirement. The balance between fused and fragmented forms of the mitochondria is governed by specific mitochondrial-shaping proteins, mitofusin 1 (Mfn1) (Santel et al. 2003), mitofusin 2 (Mfn2) (Bach et al. 2003), and optic atrophy 1 (OPA1) (Cipolat et al. 2004) for mitochondrial fusion and dynamin-related protein 1 (Drp1) (Smirnova et al. 1998) and human fission 1 (hFis1) (James et al. 2003) for mitochondrial fission. Mitochondrial fusion serves to facilitate the exchange of contents, DNA, and metabolites between neighboring mitochondria and to compensate for mutations in mtDNA (Chen et al. 2010; Nakada et al. 2001) and prevents healthy mitochondria from being removed via autophagy, a process of recycling dysfunctional mitochondria (Gomes et al. 2011). Mitochondrial fusion by inhibition of Drp1 or upregulation of Mfn1 and Mfn2 has been shown to protect the heart from ischamia–reperfusion injury via inhibition of the mitochondrial permeability transition pore (mPTP), a channel in the IMM that opens upon calcium overloading and accumulation of oxidative stress (Ong et al. 2010). Conversely, fission is crucial for enhancing optimum mitochondrial transport, to subcellular regions of specific energy demand. Fragmentation of the mitochondria by upregulation of Drp1 predisposes the cardiac cells to autophagy and cell death upon stress insults such as oxidative stress buildup and calcium overloading (Chen et al. 2005; Gomes et al. 2011; Jheng et al. 2012; Lee et al. 2011).

A variety of methods have been employed to monitor the changes in mitochondrial morphology. Traditional studies of mitochondrial morphology employ electron microscopy. However, this only provides a snapshot in time and space. Mitochondrial dynamics, which encompass the movement, change in shapes, and positioning, can be monitored using photoactivatable fluorescent proteins (mtPA-GFP). The photoactivated GFP diffuses rapidly within the entire mitochondrial matrix and can therefore be used to tag individual mitochondria (Lovy et al. 2012; Twig et al. 2006, 2008). In addition, monitoring mitochondrial fusion using the

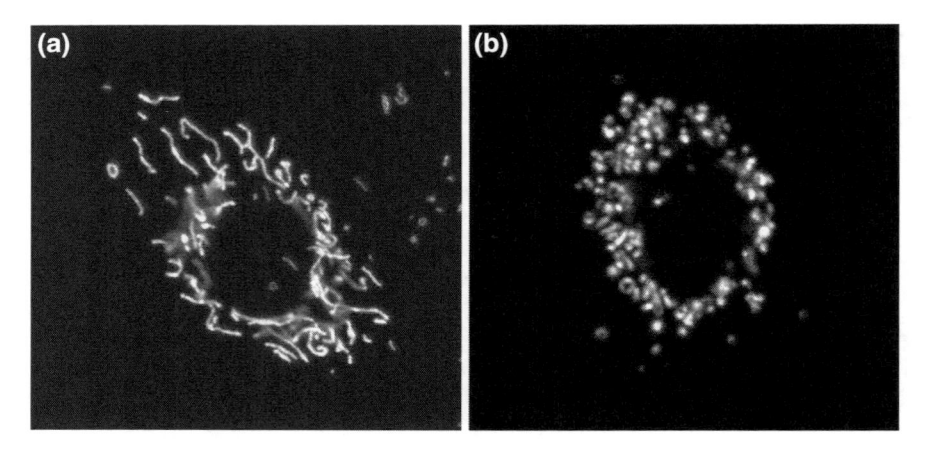

Fig. 5.1 Representative confocal microscopy images showing an endothelial cell with **a** fused mitochondria and **b** fragmented mitochondria

Fig. 5.2 Representative electron micrograph depicting an elongated mitochondrion (*in oval*) in the adult heart tissue

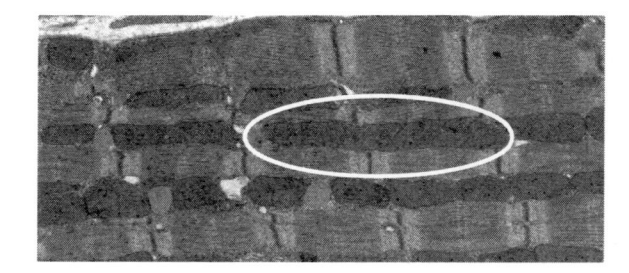

Renilla luciferase complementation assay has also been described (Huang et al. 2010; Schauss et al. 2010).

Contact between two mitochondria does not necessarily lead to fusion. Fission, conversely, can occur without movement and connection of two juxtaposed mitochondria. Fluorophores are targeted to the mitochondria via the differences in mitochondrial membrane potential to visualize the intermixing and segregation of the inner mitochondrial components. The study of mitochondrial dynamics using this tool enables a better understanding of physiological and pathophysiological processes affecting mitochondrial activity, positioning, and the number of mito-chondria (Molina and Shirihai 2009). Certain dyes such as MitoTracker® Green FM, Orange CMTMRos, Red CMXRos, and Deep Red FM probes are not affected by the energetic state of the mitochondria, compared with conventional dyes such as rhodamine 123 and tetramethylrhodamine methyl ester (TMRM). Novel, non-cytotoxic dyes, e.g., AcQCy7, have also been formulated to specifically target mitochondria (Han et al. 2013). Different model organisms, most of which express fluorescent mitochondria in vivo, have also been developed or modified to study mitochondrial dynamics in disease systems (Chandrasekaran et al. 2006; Misgeld et al. 2007; Plucińska et al. 2012). The presence of giant mitochondria can also be detected by electron microscopy (Haas et al. 2008).

The change in mitochondrial morphology has been demonstrated to affect various physiological processes. Mutations in OPA1 lead to type 1 dominant optic atrophy, a common cause of inherited visual failure starting in early childhood characterized by irreversible loss of retinal ganglion cells (RGCs) (Alavi and Fuhrmann 2013; Amati-Bonneau et al. 2009; Eiberg et al. 1994; Galvez-Ruiz et al. 2013; Kjer 1959; Lenaers et al. 2012; Lunkes et al. 1995) as well as impairment of neuronal maturation (Bertholet et al. 2013). Disruption or loss of OPA1 has been demonstrated to be associated with complex I deficiency (Ramonet et al. 2013) and Parkinson's disease (Anglade et al. 1997; Fernandes and Rao 2011; Sekiya et al. 1982; Trimmer et al. 2000) as well as affecting vertebrate development (Rahn et al. 2013). Mfn2 mutations lead to primary axonal Charcot–Marie–Tooth disease type 2A, an autosomal dominant neuropathy that impairs motor and sensory neurons with the longest axons, resulting in earliest symptoms in distal extremities (Züchner et al. 2004). Similar to OPA1, a missense mutation in Mfn2 can lead to instability of mtDNA and optic atrophy 'plus' phenotypes such as syndromic forms of autosomal dominant optic atrophy associated with sensorineural deafness, axonal sensory motor polyneuropathy, ataxia, chronic progressive external ophthalmoplegia, and mitochondrial myopathy with cytochrome c oxidase (COX)-negative and ragged-red fibers (Amati-Bonneau et al. 2008; Hudson et al. 2008; Rouzier et al. 2012). Brain MRI from the patient revealed multiple periventricular white matter lesions and severe diffuse cerebral atrophy (Rouzier et al. 2012). Fragmentation of the mitochondrial network was also observed in the fibroblasts of the patient (Rouzier et al. 2012). Concurrent with reduced levels of Mfn1, Mfn2, and OPA1 and high levels of Fis1 and Drp1, increased mitochondrial fragmentation has also been observed in cells and tissues from patients with Alzheimer's disease (Manczak et al. 2011; Wang et al. 2009). Mfn2 ablation promotes axon degeneration and disrupts axonal mitochondrial positioning in parkinsonism (Lee et al. 2012; Misko et al. 2012).

5.3.2 Determining Mitochondrial Membrane Potential

Adenosine triphosphate (ATP), the source of cellular energy from the mitochondria, is mainly produced via OXPHOS (Kaim and Dimroth 1999). The OXPHOS system is made up of five multiprotein complexes embedded within the IMM: complex I (reduced nicotinamide adenine dinucleotide: ubiquinone oxidoreductase), complex II (succinate: ubiquinone oxidoreductase), complex III (ubiquinol: cytochrome c oxidoreductase), complex IV (cytochrome c oxidase), and complex V (ATP synthase). Protons, donated by reduced nicotinamide adenine dinucleotide, ubiquinone, and cytochrome c, are channeled by complexes I, III, and IV, respectively, into the mitochondrial IMS. The channeling of protons through the complexes creates a proton gradient between the mitochondrial matrix and the

IMS. The proton gradient is crucial for generating and maintaining the electrochemical membrane potential ($\Delta\Psi$), which is the driving force for the conversion of ADP and inorganic phosphate into ATP (Kaim and Dimroth 1999).

Mitochondrial membrane potential ($\Delta\Psi_m$) can be estimated using fluorescent voltage-sensitive dyes that are membrane-permeant lipophilic cationic compounds that distribute across the IMM (Lemasters and Ramshesh 2007) based on the Nernst equation. Changes in $\Delta\Psi_m$ have been usually expressed in percentage over the basal level (Dedkova and Blatter 2012). Rhodamine 123, tetramethylrhodamine methyl, and ethyl ester (TMRM and TMRE) are specifically used to monitor changes in $\Delta\Psi_m$, which are useful for monitoring pathophysiological states or pharmacological treatment that induces changes in mitochondrial energetic states. Although it is least toxic to cells, TMRM should be used at a low concentration. An alternative fluorescent probe is the carbocyanine compound, JC-1, which exists as a green fluorescent monomer at low membrane potential but switches to form red fluorescent aggregates at higher potentials (Di Lisa et al. 1995). This change in fluorescence is solely based on membrane potential and independent of size and density of the mitochondria. Various factors such as loading times and dye concentration affect the formation of aggregates (Diaz et al. 2001; Duchen et al. 2003). Defects in complexes of the RC, which concurs with defects in OXPHOS, were detected using JC-1, as the buildup of mitochondrial membrane potential is disrupted by impairment of OXPHOS (De Paepe et al. 2012). Simultaneous detection of both $\Delta\Psi_m$ and morphology can also be achieved via a combination of potential-insensitive mitochondrial GFP and a potential-sensitive probe such as TMRM (Distelmaier et al. 2008). This will also help in correcting for movement artefacts by applying a ratiometric approach (Twig et al. 2006). Nevertheless, it is important to ascertain that certain considerations should be taken into account when using these dyes, e.g., the optimum dye concentration to be used, the presence of the dye in superfusion solution throughout the experiment, timing for addition of dye, and other factors affecting the localization of the dye such as movement and changes in volume of the mitochondria (Dedkova and Blatter 2012; Duchen et al. 2003; O'Reilly et al. 2003). The potential occurrence of FRET phenomena (Förster resonance energy transfer or fluorescence resonance energy transfer) between two different dyes that can lead to erroneous interpretations of the results should also be evaluated (Aon et al. 2007; Dedkova and Blatter 2012; Honda et al. 2005; Hüser et al. 1998; Slodzinski et al. 2008). The plasma membrane potential, together with the $\Delta\Psi_m$, dictates the accumulation of the dye (Davidson et al. 2007). Upon opening of the mPTP, dyes may leak out of the mitochondria via simple diffusion (O'Reilly et al. 2003; Twig et al. 2006). Quantitative calibration of experimental data with potentiometric probes is usually expressed as a percentage change from basal levels (Dedkova and Blatter 2012). The standard deviation (SD) of the measured fluorescence signal is important as the SD may vary from high (in polarized mitochondria) to low (in depolarized mitochondria) (Duchen et al. 2003). Calibration of $\Delta\Psi_m$ can be achieved by obtaining the ratio of mean fluorescence to SD (mean/SD), which gives a quantitative measure of dye localization (Dedkova and Blatter 2012; Duchen et al. 2003).

5.3.3 Detecting Mitochondrial Calcium Flux

Calcium (Ca^{2+}) is an important signaling molecule regulating the process of ATP synthesis and hydrolysis. In response to physiological stimuli, the concentration of mitochondrial Ca^{2+} is increased in line with the activity of mitochondrial RC to cope with enhanced energy demand and ATP synthesis (Jouaville et al. 1999; Rizzuto and Pozzan 2006). Mitochondria are crucial in maintaining intracellular calcium homeostasis. The mitochondria act as a Ca^{2+} sink to prevent propagation of large calcium waves by taking up Ca^{2+} released from the endoplasmic reticulum (ER) into the cytosol (Baughman et al. 2011; Boitier et al. 1999; Jouaville et al. 1995; Mallilankaraman et al. 2012; Plovanich et al. 2013; Tinel et al. 1999; Williams et al. 2013). The Ca^{2+} will then be released back into the cytosol by release mechanisms (Ca^{2+}/Na exchanger, Ca^{2+}/H^+ exchanger, and transient opening of the mPTP) (Ichas et al. 1997; Pinton and Rizzuto 2006; Putney and Thomas 2006; Rizzuto et al. 1993; Tian et al. 2005). Perturbations in the normal functions of cells may result in pathophysiological Ca^{2+} overloading in mitochondria, followed by the initiation of a cascade of events that lead to cell death (Hajnóczky et al. 2006). Permeabilization of the mitochondrial membrane by prolonged opening of the mPTP leads to the release of apoptogenic proteins from the IMS and subsequent cell death (Hausenloy et al. 2009; Martinou et al. 2000). Mitochondrial calcium can be measured using fluorescent Ca^{2+} indicators such as indo-1 (Miyata et al. 1991; Schreur et al. 1996), fura-2 (Abdallah et al. 2011), fluo-3 (Sedova et al. 2006), and rhod-2 (Jou et al. 1996) as well as luminescent and fluorescent targeted proteins such as aequorin, pericam, and camaleons (Bonora et al. 2013; De la Fuente et al. 2012; Fonteriz et al. 2010). The membrane-permeant fluorescent Ca^{2+} indicators [typically acetoxymethyl (AM) esters] are entrapped in the mitochondrial matrix via a de-esterification process by intracellular (cytosolic and intramitochondrial) esterases (Dedkova and Blatter 2012). Combining the use of rhod-2 with mitochondrial-targeted GFP, the mitochondrial positioning, morphology, and relation to sequestration of calcium can be concurrently studied. Potential downsides to using these dyes include the non-ratiometric nature of all rhodamine-derived indicators which may cause errors from changes in dye concentration and motion artifacts (Dedkova and Blatter 2012). Following repetitive simulation with certain agonists, Rhod-2 has been reported to induce Ca^{2+}-dependent inhibition of mitochondrial Ca^{2+} uptake (Fonteriz et al. 2010). Besides that, the optimum dye concentration has to be determined as this affects mitochondrial membrane potential, morphology, and toxin production (Fonteriz et al. 2010). Classical ratiometric indicators such as fura-2 and indo-1, conversely, have low dissociation constant (K_d) and may accumulate in the cytosol (Dedkova and Blatter 2012). Recent innovations in mitochondrial calcium measurement include the development of Mt-pericam, a mitochondrial-matrix-targeted, circularly permuted green fluorescent protein fused to calmodulin and its target peptide M13 (Nagai et al. 2001), Mitycam which is targeted to mitochondria with a standard mitochondrial-targeting sequence (subunit VIII of human cytochrome c oxidase) (Kettlewell et al. 2009; Nagai et al. 2001), a low-Ca^{2+} affinity aequorin probe which is able to measure $[Ca^{2+}]$ in the millimolar range for long period of time without problems derived from aequorin consumption (De la Fuente et al. 2012), and, most recently, GCaMP2-mt, consisting of a mitochondrial-targeting sequence attached to a high signal-to-noise Ca^{2+} sensor protein GCaMP2 to measure oxidant-induced responses of $[Ca^{2+}]$m in cultured neonatal myocytes (Iguchi et al. 2012).

5.3.4 Monitoring Mitochondrial Autophagy

Autophagy refers to the evolutionarily conserved process of segregation and recycling of damaged or unused cellular components (Klionsky 2007; Mizushima 2007). Autophagy is defined by the formation of autophagosomes—double-membrane structures that sequester defective organelles and cytotoxic protein aggregates, or even viruses or bacteria (Fig. 5.3) (Hayashi-Nishino et al. 2009; Klionsky 2007; Mizushima et al. 2011; Mizushima 2007). The autophagosomes subsequently fuse with lysosomes leading to bulk degradation of their content, and the produced nutrients will then be recycled back to the cytoplasm (Hayashi-Nishino et al. 2009; Klionsky 2007, 2008; Mizushima et al. 2011; Mizushima 2007; Ylä-Anttila et al. 2009a). Specific removal of mitochondria is termed mitophagy (Kissová et al. 2004; Rodriguez-Enriquez et al. 2004). Through mitophagy, dysfunctional mitochondria are recycled to produce useful amino acids and other nutrients.

Pathologic dysregulation of autophagy results in various disorders such as synaptic dysfunction, stroke, brain trauma, Parkinson's, Alzheimer's, Huntington's, and other neurodegenerative diseases (Chu et al. 2007; Lai et al. 2008; Liu et al. 2008; Nixon et al. 2005; Ong and Gustafsson 2012; Rudnicki et al. 2008).

Mitophagy is generally visualized via the protein LC3B, a general marker for autophagic membranes (Dagda et al. 2008; Du et al. 2009; Klionsky et al. 2012) or using electron microscopy (Klionsky et al. 2012). Autophagosomes are visualized on TEM as two parallel membrane bilayers separated by an electron-lucent cleft (Eskelinen and Kovács 2011), containing organelles that look morphologically intact, i.e., similar to the cytosol and organelles elsewhere in the cell (Ylä-Anttila et al. 2009b). Autolysosomes usually have only one limiting membrane, while the engulfed cytoplasmic material and/or organelles are at various stages of degradation (Ylä-Anttila et al. 2009b). The recommended quantification of TEM results is to obtain the volume occupied by autophagic structures and express it as a percentage of cytoplasmic or cellular volume in a predetermined number of cells (by power analysis) per sample (Klionsky et al. 2012; Sigmond et al. 2008). Autophagosomes in the brain have been evaluated using confocal immunohistochemistry and antibodies against LC3 (Lai et al. 2008). Caution should be exercised when interpreting the increase in autophagosomes as this can reflect induction of autophagy, impairment of autophagosome turnover (Kovács et al. 1986, 1987, 1988), or the inability of turnover to keep pace with increased autophagosome formation (Chu 2006).

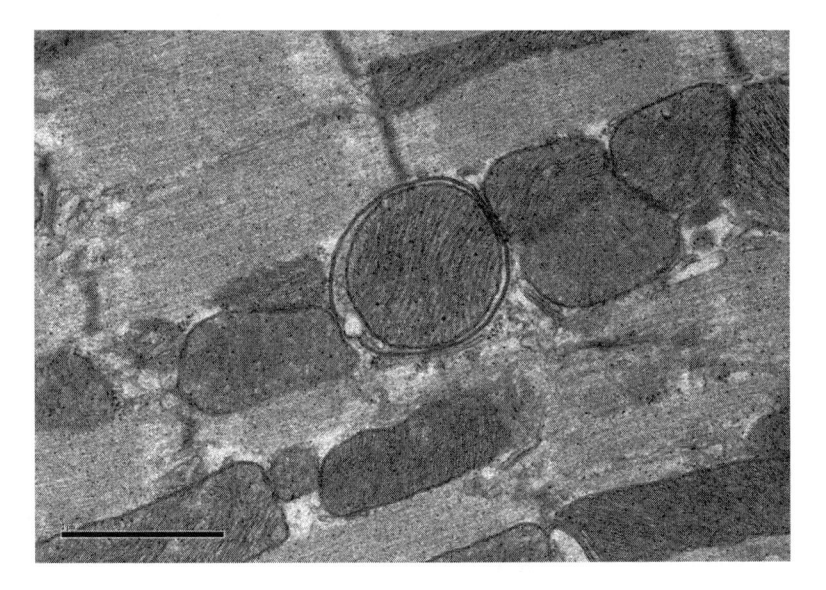

Fig. 5.3 Representative electron micrograph depicting an autophagosome with an engulfed mitochondrion in the heart

Similar to the cases described before, LC3B-GFP or LC3B-RFP can be used in conjunction with mitochondrial-targeted fluorescent proteins to concurrently study mitochondrial dynamics and mitophagy. High-quality images at a high magnification from epifluorescence or confocal laser microscope are used to quantify LC3B puncta colocalized with mitochondria, average number of LC3 puncta per cell or percentage of cells displaying punctate LC3 that exceeds a particular pre-defined threshold (Klionsky et al. 2012). Anti-LC3 antibodies for immunocytochemistry or immunohistochemistry can be used to detect the endogenous protein without the need for transfection as the overexpression of GFP-LC3 may cause the nuclear localization of the marker (Elsässer et al. 2004; Klionsky et al. 2012; Ost et al. 2010). The number of puncta corresponding to GFP-LC3 on a per cell basis rather than simply the total number of cells displaying puncta can be quantified to monitor autophagy (Klionsky et al. 2012). Sizes of autophagosomes may vary, but additional assays will be required to correlate autophagosome size with autophagic activity (Klionsky et al. 2012; Wu and Pollard 2005). When using fluorescence microscopy to monitor autophagy, a ratio for establishing differences in the degree of autophagy between cells should be obtained by calculating the SD of pixel intensities within the fluorescence image and dividing this by the mean intensity of the pixels within the area of analysis (Klionsky et al. 2012). This is crucial as the expression of GFP-LC3 may not be the same in all cells. Multispectral imaging cytometry has been proposed to assist in quantifying GFP-LC3 puncta in a large number of cells (Dolloff et al. 2011). An additional advantage of this method is the ability of quantification of endogenous LC3 in non-transfected primary cells (Phadwal et al. 2012). A caveat to using GFP-LC3 is the potential association of this chimera with protein aggregates (Kuma et al. 2007). Furthermore, using antibodies will obviate the need for generation of a transgenic organism. The late stage of mitophagy where mitochondria undergo lysosomal degradation is monitored by labeling lysosomes with lysotracker and mitochondria with MitoTracker or mitochondrial-targeted fluorescent proteins (Rodriguez-Enriquez et al. 2006). Autophagic flux can also be measured by a tandem monomeric RFP-GFP-tagged LC3 (Kimura et al. 2007) with a more sensitive and accurate reporter described in 2012 (Zhou et al. 2012). The signal elicited from mRFP is not affected by environmental conditions. The GFP signal, however, is sensitive to the acidic and/or proteolytic conditions of the lysosome lumen. Colocalization of both GFP and mRFP fluorescence indicates a phagophore or an autophagosome that has not fuse with a lysosome. In contrast, an mRFP signal without GFP may correspond to an amphisome or autolysosome. The Rosella biosensor, a dual fluorescence assay, complements the tandem mRFP/mCherry-GFP reporter by monitoring the uptake of specific cellular components to the lysosome/vacuole and the internal environment of the biosensor during autophagy (Rosado et al. 2008). The biosensor Rosella consists of a relatively pH-stable fast-maturing RFP variant and a pH-sensitive GFP variant (Rosado et al. 2008). Novel tools for studying mitochondrial turnover and biogenesis include MitoTimer, a fluorescent reporter protein that changes fluorescence from green to red concurrent with protein maturation in whole cells and isolated mitochondria (Hernandez et al. 2013), the use of Keima, a coral-derived acid-stable fluorescent protein that emits different-colored signals at acidic and neutral pHs to quantify autophagy at a single time point (Katayama et al. 2011). The correlation between component proteins and organelles involved in the timing of the autophagic process has also been studied (Karanasios et al. 2013; Katayama et al. 2011).

5.4 Conclusion

The onset of mitochondrial diseases can occur at any age with multifaceted symptoms in various organs, such as the CNS, visual system, and neuromuscular system. Diagnosis of mitochondrial diseases is not easily achievable, due to the underlying factors such as double genetic origin of RC proteins, heterogenous clinical presentations of patients, and variables between genotype and phenotype. Coupled with other investigation tools such as genetic or biochemical analysis, imaging methods are crucial to visualize and differentiate CNS abnormalities in mitochondrial disorders. Findings from the different imaging methods, however, have to be corroborated to ensure that a proper and accurate diagnosis is achieved. Advances in biomedical imaging enhance a better understanding of mitochondrial integrity and function which will help develop future therapeutics against mitochondrial diseases.

References

Abdallah Y, Kasseckert SA, Iraqi W, Said M, Shahzad T, Erdogan A, Neuhof C et al (2011) Interplay between Ca2+ cycling and mitochondrial permeability transition pores promotes reperfusion-induced injury of cardiac myocytes. J Cell Mol Med 15(11):2478–2485. doi:10.1111/j.1582-4934.2010.01249.x

Abe K (2004) Cerebral lactic acidosis correlates with neurological impairment in MELAS. Neurology 63(12):2458; author reply 2458. Retrieved from http://www.ncbi.nlm.nih.gov/pubmed/15623741

Alavi MV, Fuhrmann N (2013) Dominant optic atrophy, OPA1, and mitochondrial quality control: understanding mitochondrial network dynamics. Mol Neurodegeneration 8(1):32. doi:10.1186/1750-1326-8-32

Alpers BJ (1931) Diffuse progressive degeneration of the gray matter of the cerebrum. Arch Neurol Psychiatry 25(3):469. doi:10.1001/archneurpsyc.1931.02230030027002

Amati-Bonneau P, Valentino ML, Reynier P, Gallardo ME, Bornstein B, Boissière A, Campos Y et al (2008) OPA1 mutations induce mitochondrial DNA instability and optic atrophy "plus" phenotypes. Brain: J Neurol 131(Pt 2):338–351. doi:10.1093/brain/awm298

Amati-Bonneau P, Milea D, Bonneau D, Chevrollier A, Ferré M, Guillet V, Gueguen N et al (2009) OPA1-associated disorders: phenotypes and pathophysiology. Int J Biochem Cell Biol 41(10):1855–1865. doi:10.1016/j.biocel.2009.04.012

Anderson S, Bankier AT, Barrell BG, De Bruijn MH, Coulson AR, Drouin J, Eperon IC (1981) Sequence and organization of the human mitochondrial genome. Nature 290(5806):457–465. Retrieved from http://www.ncbi.nlm.nih.gov/pubmed/7219534

Anglade P, Vyas S, Javoy-Agid F, Herrero MT, Michel PP, Marquez J, Mouatt-Prigent A et al (1997) Apoptosis and autophagy in nigral neurons of patients with Parkinson's disease. Histol Histopathol 12(1):25–31. Retrieved from http://www.ncbi.nlm.nih.gov/pubmed/9046040

Aon MA, Cortassa S, Maack C, O'Rourke B (2007) Sequential opening of mitochondrial ion channels as a function of glutathione redox thiol status. J Biol Chem 282(30):21889–21900. doi:10.1074/jbc.M702841200

Apostolova LG, White M, Moore SA, Davis PH (2005) Deep white matter pathologic features in watershed regions: a novel pattern of central nervous system involvement in MELAS. Arch Neurol 62(7):1154–1156. doi:10.1001/archneur.62.7.1154

Ashley N, O'Rourke A, Smith C, Adams S, Gowda V, Zeviani M, Brown GK et al (2008) Depletion of mitochondrial DNA in fibroblast cultures from patients with POLG1 mutations is a consequence of catalytic mutations. Hum Mol Genet 17(16):2496–2506. doi:10.1093/hmg/ddn150

Bach D, Pich S, Soriano FX, Vega N, Baumgartner B, Oriola J, Daugaard JR et al (2003) Mitofusin-2 determines mitochondrial network architecture and mitochondrial metabolism. A novel regulatory mechanism altered in obesity. J Biol Chem 278(19):17190–17197. doi:10.1074/jbc.M212754200

Baertling F, Schaper J, Mayatepek E, Distelmaier F (2013) Teaching NeuroImages: rapidly progressive leukoencephalopathy in mitochondrial complex I deficiency. Neurology 81(2):e10–e11. doi:10.1212/WNL.0b013e31829a339b

Bardosi A, Creutzfeldt W, DiMauro S, Felgenhauer K, Friede RL, Goebel H, Kohlschütter A et al (1987) Myo-, neuro-, gastrointestinal encephalopathy (MNGIE syndrome) due to partial deficiency of cytochrome-c-oxidase. A new mitochondrial multisystem disorder. Acta Neuropathol 74(3):248–258. Retrieved from http://www.ncbi.nlm.nih.gov/pubmed/2823522

Barker PB, Hearshen DO, Boska MD (2001) Single-voxel proton MRS of the human brain at 1.5T and 3.0T. Magn Reson Med: Official J Soc Magn Reson Med/Soc Magn Reson Med 45(5):765–769. Retrieved from http://www.ncbi.nlm.nih.gov/pubmed/11323802

Barkovich AJ, Good WV, Koch TK, Berg BO (1993) Mitochondrial disorders: analysis of their clinical and imaging characteristics. AJNR Am J Neuroradiol 14(5):1119–1137. Retrieved from http://www.ncbi.nlm.nih.gov/pubmed/8237691

Barragán-Campos HM, Vallée J-N, Lô D, Barrera-Ramírez CF, Argote-Greene M, Sánchez-Guerrero J, Estañol B et al (2005) Brain magnetic resonance imaging findings in patients with mitochondrial cytopathies. Arch Neurol 62(5):737–742. doi:10.1001/archneur.62.5.737

Baughman JM, Perocchi F, Girgis HS, Plovanich M, Belcher-Timme CA, Sancak Y, Bao XR et al (2011) Integrative genomics identifies MCU as an essential component of the mitochondrial calcium uniporter. Nature 476(7360):341–345. doi:10.1038/nature10234

Bernier FP, Boneh A, Dennett X, Chow CW, Cleary MA, Thorburn DR (2002) Diagnostic criteria for respiratory chain disorders in adults and children. Neurology 59(9):1406–1411. Retrieved from http://www.ncbi.nlm.nih.gov/pubmed/12427892

Bertholet AM, Millet AME, Guillermin O, Daloyau M, Davezac N, Miquel M-C, Belenguer P (2013) OPA1 loss of function affects in vitro neuronal maturation. Brain: J Neurol 136(Pt 5):1518–1533. doi:10.1093/brain/awt060

Bianchi MC, Tosetti M, Battini R, Manca ML, Mancuso M, Cioni G, Canapicchi R et al (2003) Proton MR spectroscopy of mitochondrial diseases: analysis of brain metabolic abnormalities and their possible diagnostic relevance. AJNR Am J Neuroradiol 24(10):1958–1966. Retrieved from http://www.ncbi.nlm.nih.gov/pubmed/14625217

Boddaert N, Romano S, Funalot B, Rio M, Sarzi E, Lebre AS, Bahi-Buisson N et al (2008) 1H MRS spectroscopy evidence of cerebellar high lactate in mitochondrial respiratory chain deficiency. Mol Genet Metab 93(1):85–88. doi:10.1016/j.ymgme.2007.09.003

Boitier E, Rea R, Duchen MR (1999) Mitochondria exert a negative feedback on the propagation of intracellular Ca2+ waves in rat cortical astrocytes. J Cell Biol 145(4):795–808. Retrieved from http://www.pubmedcentral.nih.gov/articlerender.fcgi?artid=2133193&tool=pmcentrez&rendertype=abstract

Bonora M, Giorgi C, Bononi A, Marchi S, Patergnani S, Rimessi A, Rizzuto R et al (2013) Subcellular calcium measurements in mammalian cells using jellyfish photoprotein aequorin-based probes. Nat Protoc 8(11):2105–2118. doi:10.1038/nprot.2013.127

Brockmann K, Bjornstad A, Dechent P, Korenke CG, Smeitink J, Trijbels JMF, Athanassopoulos S et al (2002) Succinate in dystrophic white matter: a proton magnetic resonance spectroscopy finding characteristic for complex II deficiency. Ann Neurol 52(1):38–46. doi:10.1002/ana.10232

Burgeois M, Goutieres F, Chretien D, Rustin P, Munnich A, Aicardi J (1992) Deficiency in complex II of the respiratory chain, presenting as a leukodystrophy in two sisters with Leigh syndrome. Brain Dev 14(6):404–408. Retrieved from http://www.ncbi.nlm.nih.gov/pubmed/1492653

Cady EB, Lorek A, Penrice J, Reynolds EO, Iles RA, Burns SP, Coutts GA et al (1994) Detection of propan-1,2-diol in neonatal brain by in vivo proton magnetic resonance spectroscopy. Mag Reson Med: Official J Soc Mag Reson Med/Soc Mag Reson Med 32(6):764–767. Retrieved from http://www.ncbi.nlm.nih.gov/pubmed/7869898

Castro-Gago M, González-Conde V, Fernández-Seara MJ, Rodrigo-Sáez E, Fernández-Cebrián S, Alonso-Martín A, Campos Y et al (1999) Early mitochondrial encephalomyopathy due to complex IV deficiency consistent with Alpers-Huttenlocher syndrome: report of two cases. Rev Neurol 29(10):912–917. Retrieved from http://www.ncbi.nlm.nih.gov/pubmed/10637838

Chan DC (2006a) Mitochondria: dynamic organelles in disease, aging, and development. Cell 125(7):1241–1252. doi:10.1016/j.cell.2006.06.010

Chan DC (2006b) Mitochondrial fusion and fission in mammals. Annu Rev Cell Dev Biol 22:79–99. doi:10.1146/annurev.cellbio.22.010305.104638

Chandrasekaran K, Hazelton JL, Wang Y, Fiskum G, Kristian T (2006) Neuron-specific conditional expression of a mitochondrially targeted fluorescent protein in mice. J Neurosci: Official J Soc Neurosci 26(51):13123–13127. doi:10.1523/JNEUROSCI.4191-06.2006

Cheldi A, Ronchi D, Bordoni A, Bordo B, Lanfranconi S, Bellotti MG, Corti S et al (2013) POLG1 mutations and stroke like episodes: a distinct clinical entity rather than an atypical MELAS syndrome. BMC Neurol 13:8. doi:10.1186/1471-2377-13-8

Chen H, Chan DC (2004) Mitochondrial dynamics in mammals. Curr Top Dev Biol 59:119–144. doi:10.1016/S0070-2153(04)59005-1

Chen H, Chomyn A, Chan DC (2005) Disruption of fusion results in mitochondrial heterogeneity and dysfunction. J Biol Chem 280(28):26185–26192. doi:10.1074/jbc.m503062200

Chen H, Vermulst M, Wang YE, Chomyn A, Prolla TA, McCaffery JM, Chan DC (2010) Mitochondrial fusion is required for mtDNA stability in skeletal muscle and tolerance of mtDNA mutations. Cell 141(2):280–289. doi:10.1016/j.cell.2010.02.026

Chen C, Xiong N, Wang Y, Xiong J, Huang J, Zhang Z, Wang T (2012a) A study of familial MELAS: evaluation of A3243G mutation, clinical phenotype, and magnetic resonance spectroscopy-monitored progression. Neurol India 60(1):86–89. doi:10.4103/0028-3886.93609

Chen Z, Li J, Lou X, Ma L (2012b) Sequential evaluation of brain lesions using functional magnetic resonance imaging in patients with Leigh syndrome. Nan fang yi ke da xue xue bao = J South Med Univ 32(10):1474–1477. Retrieved from http://www.ncbi.nlm.nih.gov/pubmed/23076188

Chi C-S, Lee H-F, Tsai C-R, Chen CC-C, Tung J-N (2011a) Cranial magnetic resonance imaging findings in children with nonsyndromic mitochondrial diseases. Pediatr Neurol 44(3):171–176. doi:10.1016/j.pediatrneurol.2010.09.009

Chi C-S, Lee H-F, Tsai C-R, Chen W-S, Tung J-N, Hung H-C (2011b) Lactate peak on brain MRS in children with syndromic mitochondrial diseases. JCMA J Chin Med Assoc 74(7):305–309. doi:10.1016/j.jcma.2011.05.006

Chu CT (2006) Autophagic stress in neuronal injury and disease. J Neuropathol Exp Neurol 65(5):423–432. doi:10.1097/01.jnen.0000229233.75253.be

Chu BC, Terae S, Takahashi C, Kikuchi Y, Miyasaka K, Abe S, Minowa K et al (1999) MRI of the brain in the Kearns-Sayre syndrome: report of four cases and a review. Neuroradiology 41(10):759–764. Retrieved from http://www.ncbi.nlm.nih.gov/pubmed/10552027

Chu CT, Zhu J, Dagda R (2007) Beclin 1-independent pathway of damage-induced mitophagy and autophagic stress: implications for neurodegeneration and cell death. Autophagy 3(6):663–666. Retrieved from http://www.pubmedcentral.nih.gov/articlerender.fcgi?artid=2779565&tool=pmcentrez&rendertype=abstract

Cipolat S, Martins de Brito O, Dal Zilio B, Scorrano L (2004) OPA1 requires mitofusin 1 to promote mitochondrial fusion. Proc Natl Acad Sci USA 101(45):15927–15932. doi:10.1073/pnas.0407043101

Conway LJ, Robertson TE, McGill JJ, Hanson JP (2011) MELAS syndrome in an Indigenous Australian woman. Med J Aust 195(10):581–582. Retrieved from http://www.ncbi.nlm.nih.gov/pubmed/22107001

Dagda RK, Zhu J, Kulich SM, Chu CT (2008) Mitochondrially localized ERK2 regulates mitophagy and autophagic cell stress: implications for Parkinson's disease. Autophagy 4(6):770–782. Retrieved from http://www.pubmedcentral.nih.gov/articlerender.fcgi?artid=2574804&tool=pmcentrez&rendertype=abstract

Davidson SM, Duchen MR (2012) Imaging mitochondrial calcium signalling with fluorescent probes and single or two photon confocal microscopy. Methods Mol Biol (Clifton, NJ) 810:219–234. doi:10.1007/978-1-61779-382-0_14

Davidson SM, Yellon D, Duchen MR (2007) Assessing mitochondrial potential, calcium, and redox state in isolated mammalian cells using confocal microscopy. Methods Mol Biol (Clifton, NJ) 372:421–430. doi:10.1007/978-1-59745-365-3_30

De la Fuente S, Fonteriz RI, De la Cruz PJ, Montero M, Alvarez J (2012) Mitochondrial free [Ca(2+)] dynamics measured with a novel low-Ca(2+) affinity aequorin probe. Biochem J 445(3):371–376. doi:10.1042/BJ20120423

De Paepe B, Smet J, Vanlander A, Seneca S, Lissens W, De Meirleir L, Vandewoestyne M et al (2012) Fluorescence imaging of mitochondria in cultured skin fibroblasts: a useful method for the detection of oxidative phosphorylation defects. Pediatr Res 72(3):232–240. doi:10.1038/pr.2012.84

Dedkova EN, Blatter LA (2012) Measuring mitochondrial function in intact cardiac myocytes. J Mol Cell Cardiol 52(1):48–61. doi:10.1016/j.yjmcc.2011.08.030

Delonlay P, Rötig A, Sarnat HB (2013) Respiratory chain deficiencies. Handb Clin Neurol 113:1651–1666. doi:10.1016/B978-0-444-59565-2.00033-2

Deschauer M, Tennant S, Rokicka A, He L, Kraya T, Turnbull DM, Zierz S et al (2007) MELAS associated with mutations in the POLG1 gene. Neurology 68(20):1741–1742. doi:10.1212/01. wnl.0000261929.92478.3e

Di Lisa F, Blank PS, Colonna R, Gambassi G, Silverman HS, Stern MD, Hansford RG (1995) Mitochondrial membrane potential in single living adult rat cardiac myocytes exposed to anoxia or metabolic inhibition. J Physiol 486 (Pt 1):1–13. Retrieved from http://www.pubmedcentral.nih.gov/articlerender.fcgi?artid=1156492&tool=pmcentrez&rendertype=abstract

Diaz G, Diana A, Falchi AM, Gremo F, Pani A, Batetta B, Dessì S et al (2001) Intra- and intercellular distribution of mitochondrial probes and changes after treatment with MDR modulators. IUBMB Life 51(2):121–126. doi:10.1080/15216540119470

DiMauro S (2004) Mitochondrial diseases. Biochim Biophys Acta 1658(1–2):80–88. doi:10.1016/j.bbabio.2004.03.014

DiMauro S, Hirano M (2005) Mitochondrial encephalomyopathies: an update. Neuromuscul Disord: NMD 15(4):276–286. doi:10.1016/j.nmd.2004.12.008

DiMauro S, Schon EA (2003) Mitochondrial respiratory-chain diseases. N Engl J Med 348(26):2656–2668. doi:10.1056/NEJMra022567

DiMauro S, Bonilla E, Zeviani M, Nakagawa M, DeVivo DC (1985) Mitochondrial myopathies. Ann Neurol 17(6):521–538. doi:10.1002/ana.410170602

DiMauro S, Bonilla E, Lombes A, Shanske S, Minetti C, Moraes CT (1990) Mitochondrial encephalomyopathies. Neurologic clinics 8(3):483–506. Retrieved from http://www.ncbi.nlm.nih.gov/pubmed/2170831

Dinopoulos A, Cecil KM, Schapiro MB, Papadimitriou A, Hadjigeorgiou GM, Wong B, de Grauw T et al (2005) Brain MRI and proton MRS findings in infants and children with respiratory chain defects. Neuropediatrics 36(5):290–301. doi:10.1055/s-2005-872807

Distelmaier F, Koopman WJH, Testa ER, De Jong AS, Swarts HG, Mayatepek E, Smeitink JAM et al (2008) Life cell quantification of mitochondrial membrane potential at the single organelle level. Cytometry Part A: J Int Soc Anal Cytol 73(2):129–138. doi:10.1002/cyto.a.20503

Dolloff NG, Ma X, Dicker DT, Humphreys RC, Li LZ, El-Deiry WS (2011) Spectral imaging-based methods for quantifying autophagy and apoptosis. Cancer Biol Ther 12(4):349–356. Retrieved from http://www.pubmedcentral.nih.gov/articlerender.fcgi?artid=3230317&tool=pmcentrez&rendertype=abstract

Du L, Hickey RW, Bayir H, Watkins SC, Tyurin VA, Guo F, Kochanek PM et al (2009) Starving neurons show sex difference in autophagy. J Biol Chem 284(4):2383–2396. doi:10.1074/jbc.M804396200

Duchen MR, Surin A, Jacobson J (2003) Imaging mitochondrial function in intact cells. Methods Enzymol 361:353–389. Retrieved from http://www.ncbi.nlm.nih.gov/pubmed/12624920

Duning T, Deppe M, Keller S, Mohammadi S, Schiffbauer H, Marziniak M (2009) Diffusion tensor imaging in a case of Kearns-Sayre syndrome: striking brainstem involvement as a possible cause of oculomotor symptoms. J Neurol Sci 281(1–2):110–112. doi:10.1016/j.jns.2009.03.007

Eiberg H, Kjer B, Kjer P, Rosenberg T (1994) Dominant optic atrophy (OPA1) mapped to chromosome 3q region. I. Linkage analysis. Hum Mol Genet 3(6):977–980. Retrieved from http://www.ncbi.nlm.nih.gov/pubmed/7951248

Elsässer A, Vogt AM, Nef H, Kostin S, Möllmann H, Skwara W, Bode C et al (2004) Human hibernating myocardium is jeopardized by apoptotic and autophagic cell death. J Am Coll Cardiol 43(12):2191–2199. doi:10.1016/j.jacc.2004.02.053

Emmanuele V, Sotiriou E, Rios PG, Ganesh J, Ichord R, Foley AR, Akman HO et al (2013) A novel mutation in the mitochondrial DNA cytochrome b gene (MTCYB) in a patient with mitochondrial encephalomyopathy, lactic acidosis, and strokelike episodes syndrome. J Child Neurol 28(2):236–242. doi:10.1177/0883073812445787

Eskelinen E-L, Kovács AL (2011) Double membranes vs. lipid bilayers, and their significance for correct identification of macroautophagic structures. Autophagy 7(9):931–932. Retrieved from http://www.ncbi.nlm.nih.gov/pubmed/21642767

Farina L, Chiapparini L, Uziel G, Bugiani M, Zeviani M, Savoiardo M (2002) MR findings in Leigh syndrome with COX deficiency and SURF-1 mutations. AJNR Am J Neuroradiol 23(7):1095–1100. Retrieved from http://www.ncbi.nlm.nih.gov/pubmed/12169463

Fernandes C, Rao Y (2011) Genome-wide screen for modifiers of Parkinson's disease genes in Drosophila. Mol brain 4:17. doi:10.1186/1756-6606-4-17

Finsterer J (2009) Central nervous system imaging in mitochondrial disorders. Can J Neurol Sci. Le journal canadien des sciences neurologiques 36(2):143–153. Retrieved from http://www.ncbi.nlm.nih.gov/pubmed/19378706

Finsterer J, Jarius C, Eichberger H (2001) Phenotype variability in 130 adult patients with respiratory chain disorders. J Inherit Metab Dis 24(5):560–576. Retrieved from http://www.ncbi.nlm.nih.gov/pubmed/11757584

Fonteriz RI, De la Fuente S, Moreno A, Lobatón CD, Montero M, Alvarez J (2010) Monitoring mitochondrial [Ca(2+)] dynamics with rhod-2, ratiometric pericam and aequorin. Cell Calcium 48(1):61–69. doi:10.1016/j.ceca.2010.07.001

Friedman SD, Shaw DWW, Ishak G, Gropman AL, Saneto RP (2010) The use of neuroimaging in the diagnosis of mitochondrial disease. Dev Disabil Res Rev 16(2):129–135. doi:10.1002/ddrr.103

Galluzzi L, Kepp O, Trojel-Hansen C, Kroemer G (2012) Mitochondrial control of cellular life, stress, and death. Circ Res 111(9):1198–1207. doi:10.1161/CIRCRESAHA.112.268946

Galvez-Ruiz A, Neuhaus C, Bergmann C, Bolz H (2013) First cases of dominant optic atrophy in Saudi Arabia: report of two novel OPA1 mutations. J Neuroophthalmol: Official J N Am Neuroophthalmol Soc. doi:10.1097/WNO.0b013e31829ffb9a

Gieraerts C, Demaerel P, Van Damme P, Wilms G (2013) Mitochondrial encephalomyopathy, lactic acidosis, and stroke-like episodes (MELAS) syndrome mimicking herpes simplex encephalitis on imaging studies. J Comput Assist Tomogr 37(2):279–281. doi:10.1097/RCT.0b013e3182811170

Gire C, Girard N, Nicaise C, Einaudi MA, Montfort MF, Dejode JM (2002) Clinical features and neuroradiological findings of mitochondrial pathology in six neonates. Child's Nerv Syst: ChNS: Official J Int Soc Pediatr Neurosurg 18(11):621–628. doi:10.1007/s00381-002-0621-0

Goethem Van, Gert Schwartz M, Löfgren A, Dermaut B, Van Broeckhoven C, Vissing J (2003) Novel POLG mutations in progressive external ophthalmoplegia mimicking mitochondrial neurogastrointestinal encephalomyopathy. EJHG Eur J Hum Genet 11(7):547–549. doi:10.1038/sj.ejhg.5201002

Gomes LC, Di Benedetto G, Scorrano L (2011) During autophagy mitochondria elongate, are spared from degradation and sustain cell viability. Nat Cell Biol 13(5):589–598. doi:10.1038/ncb2220

Goto Y, Nonaka I, Horai S (1990) A mutation in the tRNA(Leu)(UUR) gene associated with the MELAS subgroup of mitochondrial encephalomyopathies. Nature 348(6302):651–653. doi:10.1038/348651a0

Haas R, Dietrich R (2004) Neuroimaging of mitochondrial disorders. Mitochondrion 4(5–6):471–490. doi:10.1016/j.mito.2004.07.008

Haas RH, Parikh S, Falk MJ, Saneto RP, Wolf NI, Darin N, Cohen BH (2007) Mitochondrial disease: a practical approach for primary care physicians. Pediatrics 120(6):1326–1333. doi:10.1542/peds.2007-0391

Haas RH, Parikh S, Falk MJ, Saneto RP, Wolf NI, Darin N, Wong L-J et al (2008) The in-depth evaluation of suspected mitochondrial disease. Mol Genet Metab 94(1):16–37. doi:10.1016/j.ymgme.2007.11.018

Hajnóczky G, Csordás G, Das S, Garcia-Perez C, Saotome M, Sinha Roy S, Yi M (2006) Mitochondrial calcium signalling and cell death: approaches for assessing the role of mitochondrial Ca2+ uptake in apoptosis. Cell Calcium 40(5–6):553–560. doi:10.1016/j.ceca.2006.08.016

Hakonen AH, Isohanni P, Paetau A, Herva R, Suomalainen A, Lönnqvist T (2007) Recessive twinkle mutations in early onset encephalopathy with mtDNA depletion. Brain: J Neurol 130(Pt 11):3032–3040. doi:10.1093/brain/awm242

Hall AM, Unwin RJ, Parker N, Duchen MR (2009) Multiphoton imaging reveals differences in mitochondrial function between nephron segments. JASN J Am Soc Nephrol 20(6):1293–1302. doi:10.1681/ASN.2008070759

Hall AM, Rhodes GJ, Sandoval RM, Corridon PR, Molitoris BA (2013) In vivo multiphoton imaging of mitochondrial structure and function during acute kidney injury. Kidney Int 83(1):72–83. doi:10.1038/ki.2012.328

Han J, Han MS, Tung C-H (2013) A non-toxic fluorogenic dye for mitochondria labeling. Biochim Biophys Acta 1830(11):5130–5135. doi:10.1016/j.bbagen.2013.07.001

Harding BN (1990) Progressive neuronal degeneration of childhood with liver disease (Alpers-Huttenlocher syndrome): a personal review. J Child Neurol 5(4):273–287. Retrieved from http://www.ncbi.nlm.nih.gov/pubmed/2246481

Hausenloy DJ, Ong S-B, Yellon DM (2009) The mitochondrial permeability transition pore as a target for preconditioning and postconditioning. Basic Res Cardiol 104(2):189–202. doi:10.1007/s00395-009-0010-x

Hayashi-Nishino M, Fujita N, Noda T, Yamaguchi A, Yoshimori T, Yamamoto A (2009) A subdomain of the endoplasmic reticulum forms a cradle for autophagosome formation. Nat Cell Biol 11(12):1433–1437. doi:10.1038/ncb1991

Hernandez G, Thornton C, Stotland A, Lui D, Sin J, Ramil J, Magee N et al (2013) MitoTimer: a novel tool for monitoring mitochondrial turnover. Autophagy 9(11):1852–1861. Retrieved from http://www.ncbi.nlm.nih.gov/pubmed/24128932

Hirano M, Pavlakis SG (1994) Mitochondrial myopathy, encephalopathy, lactic acidosis, and strokelike episodes (MELAS): current concepts. J Child Neurol 9(1):4–13. Retrieved from http://www.ncbi.nlm.nih.gov/pubmed/8151079

Honda HM, Korge P, Weiss JN (2005) Mitochondria and ischemia/reperfusion injury. Ann N Y Acad Sci 1047:248–258. doi:10.1196/annals.1341.022

Horvath R, Hudson G, Ferrari G, Fütterer N, Ahola S, Lamantea E, Prokisch H et al (2006) Phenotypic spectrum associated with mutations of the mitochondrial polymerase gamma gene. Brain: J Neurol 129(Pt 7):1674–1684. doi:10.1093/brain/awl088

Hourani RG, Barada WM, Al-Kutoubi AM, Hourani MH (2006) Atypical MRI findings in Kearns-Sayre syndrome: T2 radial stripes. Neuropediatrics 37(2):110–113. doi:10.1055/s-2006-924226

Huang H, Choi S-Y, Frohman MA (2010) A quantitative assay for mitochondrial fusion using Renilla luciferase complementation. Mitochondrion 10(5):559–566. doi:10.1016/j.mito.2010.05.003

Hudson G, Amati-Bonneau P, Blakely EL, Stewart JD, He L, Schaefer AM, Griffiths PG et al (2008) Mutation of OPA1 causes dominant optic atrophy with external ophthalmoplegia, ataxia, deafness and multiple mitochondrial DNA deletions: a novel disorder of mtDNA maintenance. Brain: J Neurol 131(Pt 2):329–337. doi:10.1093/brain/awm272

Hunter MF, Peters H, Salemi R, Thorburn D, Mackay MT (2011) Alpers syndrome with mutations in POLG: clinical and investigative features. Pediatr Neurol 45(5):311–318. doi:10.1016/j.pediatrneurol.2011.07.008

Hüser J, Rechenmacher CE, Blatter LA (1998) Imaging the permeability pore transition in single mitochondria. Biophys J 74(4):2129–2137. doi:10.1016/S0006-3495(98)77920-2

Huttenlocher PR, Solitare GB, Adams G (1976) Infantile diffuse cerebral degeneration with hepatic cirrhosis. Arch Neurol 33(3):186–192. Retrieved from http://www.ncbi.nlm.nih.gov/pubmed/1252162

Ichas F, Jouaville LS, Mazat JP (1997) Mitochondria are excitable organelles capable of generating and conveying electrical and calcium signals. Cell 89(7):1145–1153. Retrieved from http://www.ncbi.nlm.nih.gov/pubmed/9215636

Iguchi M, Kato M, Nakai J, Takeda T, Matsumoto-Ida M, Kita T, Kimura T et al (2012) Direct monitoring of mitochondrial calcium levels in cultured cardiac myocytes using a novel fluorescent indicator protein, GCaMP2-mt. Int J Cardiol 158(2):225–234. doi:10.1016/j.ijcard. 2011.01.034

Iizuka T, Sakai F, Suzuki N, Hata T, Tsukahara S, Fukuda M, Takiyama Y (2002) Neuronal hyperexcitability in stroke-like episodes of MELAS syndrome. Neurology 59(6):816–824. Retrieved from http://www.ncbi.nlm.nih.gov/pubmed/12297560

Iizuka T, Sakai F, Ide T, Miyakawa S, Sato M, Yoshii S (2007) Regional cerebral blood flow and cerebrovascular reactivity during chronic stage of stroke-like episodes in MELAS— implication of neurovascular cellular mechanism. J Neurol Sci 257(1–2):126–138. doi:10. 1016/j.jns.2007.01.040

Ingman M, Kaessmann H, Pääbo S, Gyllensten U (2000) Mitochondrial genome variation and the origin of modern humans. Nature 408(6813):708–713. doi:10.1038/35047064

Ionasescu V (1983) Oculogastrointestinal muscular dystrophy. Am J Med Genet 15(1):103–112. doi:10.1002/ajmg.1320150114

Isobe T, Matsumura A, Anno I, Kawamura H, Shibata Y, Muraishi H, Minami M (2007) Lactate quantification by proton magnetic resonance spectroscopy using a clinical MRI machine: a basic study. Australas Radiol 51(4):330–333. doi:10.1111/j.1440-1673.2007.01745.x

Jain-Ghai S, Cameron JM, Al Maawali A, Blaser S, MacKay N, Robinson B, Raiman J (2013) Complex II deficiency–a case report and review of the literature. Am J Med Genet. Part A 161A(2):285–294. doi:10.1002/ajmg.a.35714

Jaksch M, Horvath R, Horn N, Auer DP, Macmillan C, Peters J, Gerbitz KD et al (2001) Homozygosity (E140K) in SCO$_2$ causes delayed infantile onset of cardiomyopathy and neuropathy. Neurology 57(8):1440–1146. Retrieved from http://www.ncbi.nlm.nih.gov/ pubmed/11673586

James DI, Parone PA, Mattenberger Y, Martinou J-C (2003) hFis1, a novel component of the mammalian mitochondrial fission machinery. J Biol Chem 278(38):36373–36379. doi:10. 1074/jbc.M303758200

Janssen RJRJ, Nijtmans LG, Van den Heuvel LP, Smeitink JAM (2006) Mitochondrial complex I: structure, function and pathology. J Inherit Metab Dis 29(4):499–515. doi:10.1007/ s10545-006-0362-4

Jeppesen TD, Schwartz M, Hansen K, Danielsen ER, Wibrand F, Vissing J (2003) Late onset of stroke-like episode associated with a 3256C–> T point mutation of mitochondrial DNA. J Neurol Sci 214(1–2):17–20. Retrieved from http://www.ncbi.nlm.nih.gov/pubmed/ 12972383

Jheng H-F, Tsai P-J, Guo S-M, Kuo L-H, Chang C-S, Su I-J, Chang C-R et al (2012) Mitochondrial fission contributes to mitochondrial dysfunction and insulin resistance in skeletal muscle. Mol Cell Biol 32(2):309–319. doi:10.1128/MCB.05603-11

José da Rocha A, Túlio Braga F, Carlos Martins Maia A, Jorge da Silva C, Toyama C, Pereira Pinto Gama H, Kok F et al (2008) Lactate detection by MRS in mitochondrial encephalopathy: optimization of technical parameters. J Neuroimaging: Official J Am Soc Neuroimaging 18(1):1–8. doi:10.1111/j.1552-6569.2007.00205.x

Jou MJ, Peng TI, Sheu SS (1996) Histamine induces oscillations of mitochondrial free Ca2+ concentration in single cultured rat brain astrocytes. J Physiol 497(Pt 2):299–308. Retrieved from http://www.pubmedcentral.nih.gov/articlerender.fcgi?artid=1160985&tool=pmcentrez& rendertype=abstract

Jouaville LS, Ichas F, Holmuhamedov EL, Camacho P, Lechleiter JD (1995) Synchronization of calcium waves by mitochondrial substrates in Xenopus laevis oocytes. Nature 377(6548):438–441. doi:10.1038/377438a0

Jouaville LS, Pinton P, Bastianutto C, Rutter GA, Rizzuto R (1999) Regulation of mitochondrial ATP synthesis by calcium: evidence for a long-term metabolic priming. Proc Nat Acad Sci USA 96(24):13807–13812. Retrieved from http://www.pubmedcentral.nih.gov/articlerender. fcgi?artid=24146&tool=pmcentrez&rendertype=abstract

Kaim G, Dimroth P (1999) ATP synthesis by F-type ATP synthase is obligatorily dependent on the transmembrane voltage. EMBO J 18(15):4118–4127. doi:10.1093/emboj/18.15.4118

Kamata Y, Mashima Y, Yokoyama M, Tanaka K, Goto Y, Oguchi Y (1998) Patient with Kearns-Sayre syndrome exhibiting abnormal magnetic resonance image of the brain. J Neuroophthalmol: Official J North Am Neuroophthalmol Soc 18(4):284–288. Retrieved from http://www.ncbi.nlm.nih.gov/pubmed/9858014

Karanasios E, Stapleton E, Walker SA, Manifava M, Ktistakis NT (2013) Live cell imaging of early autophagy events: omegasomes and beyond. JoVE J Visualized Exp (77). doi:10.3791/50484

Karbowski M, Arnoult D, Chen H, Chan DC, Smith CL, Youle RJ (2004) Quantitation of mitochondrial dynamics by photolabeling of individual organelles shows that mitochondrial fusion is blocked during the Bax activation phase of apoptosis. J Cell Biol 164(4):493–499. doi:10.1083/jcb.200309082

Katayama H, Kogure T, Mizushima N, Yoshimori T, Miyawaki A (2011) A sensitive and quantitative technique for detecting autophagic events based on lysosomal delivery. Chem Biol 18(8):1042–1052. doi:10.1016/j.chembiol.2011.05.013

Kearns TP, Sayre GP (1958) Retinitis pigmentosa, external ophthalmoplegia, and complete heart block: unusual syndrome with histologic study in one of two cases. AMA Arch Ophthalmol 60(2):280–289. Retrieved from http://www.ncbi.nlm.nih.gov/pubmed/13558799

Keevil SF (2006) Spatial localization in nuclear magnetic resonance spectroscopy. Phys Med Biol 51(16):R579–R636. doi:10.1088/0031-9155/51/16/R01

Kettlewell S, Cabrero P, Nicklin SA, Dow JAT, Davies S, Smith GL (2009) Changes of intra-mitochondrial Ca2+ in adult ventricular cardiomyocytes examined using a novel fluorescent Ca2+ indicator targeted to mitochondria. J Mol Cell Cardiol 46(6):891–901. doi:10.1016/j.yjmcc.2009.02.016

Kim HS, Kim DI, Lee BI, Jeong EK, Choi C, Lee JD, Yoon PH et al (2001) Diffusion-weighted image and MR spectroscopic analysis of a case of MELAS with repeated attacks. Yonsei Med J 42(1):128–133. Retrieved from http://www.ncbi.nlm.nih.gov/pubmed/11293491

Kimura S, Noda T, Yoshimori T (2007) Dissection of the autophagosome maturation process by a novel reporter protein, tandem fluorescent-tagged LC3. Autophagy 3(5):452–460. Retrieved from http://www.ncbi.nlm.nih.gov/pubmed/17534139

Kissová I, Deffieu M, Manon S, Camougrand N (2004) Uth1p is involved in the autophagic degradation of mitochondria. J Biol Chem 279(37):39068–39074. doi:10.1074/jbc.M406960200

Kjer P (1959) Infantile optic atrophy with dominant mode of inheritance: a clinical and genetic study of 19 Danish families. Acta Ophthalmol Suppl 164(Supp 54):1–147. Retrieved from http://www.ncbi.nlm.nih.gov/pubmed/13660776

Klionsky DJ (2007) Autophagy: from phenomenology to molecular understanding in less than a decade. Nat Rev Mol Cell Biol 8(11):931–937. doi:10.1038/nrm2245

Klionsky DJ (2008) Autophagy revisited: a conversation with Christian de Duve. Autophagy 4(6):740–743. Retrieved from http://www.ncbi.nlm.nih.gov/pubmed/18567941

Klionsky DJ, Abdalla FC, Abeliovich H, Abraham RT, Acevedo-Arozena A, Adeli K, Agholme L et al (2012) Guidelines for the use and interpretation of assays for monitoring autophagy. Autophagy 8(4):445–544

Knight MA, Gardner RJM, Bahlo M, Matsuura T, Dixon JA, Forrest SM, Storey E (2004) Dominantly inherited ataxia and dysphonia with dentate calcification: spinocerebellar ataxia type 20. Brain: J Neurol 127(Pt 5):1172–1181. doi:10.1093/brain/awh139

Kovács J, Fellinger E, Kárpáti PA, Kovács AL, László L (1986) The turnover of autophagic vacuoles: evaluation by quantitative electron microscopy. Biomed Biochim Acta 45(11–12):1543–1547. Retrieved from http://www.ncbi.nlm.nih.gov/pubmed/3579875

Kovács J, Fellinger E, Kárpáti AP, Kovács AL, László L, Réz G (1987) Morphometric evaluation of the turnover of autophagic vacuoles after treatment with Triton X-100 and vinblastine in murine pancreatic acinar and seminal vesicle epithelial cells. Virchows Archiv B, Cell Pathol

Incl Mol Pathol 53(3):183–190. Retrieved from http://www.ncbi.nlm.nih.gov/pubmed/2888237

Kovács J, László L, Kovács AL (1988) Regression of autophagic vacuoles in pancreatic acinar, seminal vesicle epithelial, and liver parenchymal cells: a comparative morphometric study of the effect of vinblastine and leupeptin followed by cycloheximide treatment. Exp Cell Res 174(1):244–251. Retrieved from http://www.ncbi.nlm.nih.gov/pubmed/3335225

Kuma A, Matsui M, Mizushima N (2007) LC3, an autophagosome marker, can be incorporated into protein aggregates independent of autophagy: caution in the interpretation of LC3 localization. Autophagy 3(4):323–328. Retrieved from http://www.ncbi.nlm.nih.gov/pubmed/17387262

Kurt B, Jaeken J, Van Hove J, Lagae L, Löfgren A, Everman DB, Jayakar P et al (2010) A novel POLG gene mutation in 4 children with Alpers-like hepatocerebral syndromes. Arch Neurol 67(2):239–244. doi:10.1001/archneurol.2009.332

Lai Y, Hickey RW, Chen Y, Bayir H, Sullivan ML, Chu CT, Kochanek PM et al (2008) Autophagy is increased after traumatic brain injury in mice and is partially inhibited by the antioxidant gamma-glutamylcysteinyl ethyl ester. J Cereb Blood Flow Metab: Official J Int Soc Cereb Blood Flow Metab 28(3):540–550. doi:10.1038/sj.jcbfm.9600551

Lamperti C, Diodato D, Lamantea E, Carrara F, Ghezzi D, Mereghetti P, Rizzi R et al (2012) MELAS-like encephalomyopathy caused by a new pathogenic mutation in the mitochondrial DNA encoded cytochrome c oxidase subunit I. Neuromuscul Disord. Retrieved from http://www.sciencedirect.com/science/article/pii/S096089661200185X

Lange T, Dydak U, Roberts TPL, Rowley HA, Bjeljac M, Boesiger P (2006) Pitfalls in lactate measurements at 3T. AJNR Am J Neuroradiol 27(4):895–901. Retrieved from http://www.ncbi.nlm.nih.gov/pubmed/16611787

Larsson NG, Holme E, Kristiansson B, Oldfors A, Tulinius M (1990) Progressive increase of the mutated mitochondrial DNA fraction in Kearns-Sayre syndrome. Pediatr Res 28(2):131–6. Retrieved from http://www.ncbi.nlm.nih.gov/pubmed/2395603

Lebre AS, Rio M, Faivre d'Arcier L, Vernerey D, Landrieu P, Slama A, Jardel C et al (2011) A common pattern of brain MRI imaging in mitochondrial diseases with complex I deficiency. J Med Genet 48(1):16–23. doi:10.1136/jmg.2010.079624

Lee H-F, Lee H-J, Chi C-S, Tsai C-R, Chang T-K, Wang C-J (2007) The neurological evolution of Pearson syndrome: case report and literature review. Eur J Paediatr Neurol: EJPN: Official J Eur Paediatr Neurol Soc 11(4):208–214. doi:10.1016/j.ejpn.2006.12.008

Lee Y, Lee H-Y, Hanna RA, Gustafsson ÅB (2011) Mitochondrial autophagy by Bnip3 involves Drp1-mediated mitochondrial fission and recruitment of Parkin in cardiac myocytes. Am J Physiol Heart Circ Physiol 301(5):H1924–H1931. doi:10.1152/ajpheart.0 0368.2011

Lee S, Sterky FH, Mourier A, Terzioglu M, Cullheim S, Olson L, Larsson N-G (2012) Mitofusin 2 is necessary for striatal axonal projections of midbrain dopamine neurons. Hum Mol Genet 21(22):4827–4835. doi:10.1093/hmg/dds352

Leigh D (1951) Subacute necrotizing encephalomyelopathy in an infant. J Neurol, Neurosurg, Psychiatry 14(3):216–221. Retrieved from http://www.pubmedcentral.nih.gov/articlerender.fcgi?artid=499520&tool=pmcentrez&rendertype=abstract

Lemasters JJ, Ramshesh VK (2007) Imaging of mitochondrial polarization and depolarization with cationic fluorophores. Methods Cell Biol 80:283–295. doi:10.1016/S0091-679X(06)80014-2

Lenaers G, Hamel C, Delettre C, Amati-Bonneau P, Procaccio V, Bonneau D, Reynier P et al (2012) Dominant optic atrophy. Orphanet J Rare Dis 7:46. doi:10.1186/1750-1172-7-46

Lerman-Sagie T, Leshinsky-Silver E, Watemberg N, Luckman Y, Lev D (2005) White matter involvement in mitochondrial diseases. Mol Genet Metab 84(2):127–136. doi:10.1016/j.ymgme.2004.09.008

Leutner C, Layer G, Zierz S, Solymosi L, Dewes W, Reiser M (1994) Cerebral MR in ophthalmoplegia plus. AJNR Am J Neuroradiol 15(4):681–687. Retrieved from http://www.ncbi.nlm.nih.gov/pubmed/8010270

Libernini L, Lupis C, Mastrangelo M, Carrozzo R, Santorelli FM, Inghilleri M, Leuzzi V (2012) Mitochondrial neurogastrointestinal encephalomyopathy: novel pathogenic mutations in thymidine phosphorylase gene in two Italian brothers. Neuropediatrics 43(4):201–208. doi:10.1055/s-0032-1315431

Liesa M, Palacín M, Zorzano A (2009) Mitochondrial dynamics in mammalian health and disease. Physiol Rev 89(3):799–845. doi:10.1152/physrev.0 0030.2008

Lin DDM, Crawford TO, Barker PB (2003) Proton MR spectroscopy in the diagnostic evaluation of suspected mitochondrial disease. AJNR Am J Neuroradiol 24(1):33–41. Retrieved from http://www.ncbi.nlm.nih.gov/pubmed/12533324

Liu CL, Chen S, Dietrich D, Hu BR (2008) Changes in autophagy after traumatic brain injury. J Cereb Blood Flow Metab: Official J Int Soc Cereb Blood Flow Metab 28(4):674–683. doi:10.1038/sj.jcbfm.9600587

Lovy A, Molina AJA, Cerqueira FM, Trudeau K, Shirihai OS (2012) A faster, high resolution, mtPA-GFP-based mitochondrial fusion assay acquiring kinetic data of multiple cells in parallel using confocal microscopy. JoVE J Visualized Exp 65:e3991. doi:10.3791/3991

Lunkes A, Hartung U, Magariño C, Rodríguez M, Palmero A, Rodríguez L, Heredero L et al (1995) Refinement of the OPA1 gene locus on chromosome 3q28-q29 to a region of 2–8 cM, in one Cuban pedigree with autosomal dominant optic atrophy type Kjer. Am J Hum Genet 57(4):968–970. Retrieved from http://www.pubmedcentral.nih.gov/articlerender.fcgi?artid=1801490&tool=pmcentrez&rendertype=abstract

Majoie CB, Akkerman EM, Blank C, Barth PG, Poll-The BT, Den Heeten GJ (2002) Mitochondrial encephalomyopathy: comparison of conventional MR imaging with diffusion-weighted and diffusion tensor imaging: case report. AJNR Am J Neuroradiol 23(5):813–816. Retrieved from http://www.ncbi.nlm.nih.gov/pubmed/12006283

Malfatti E, Bugiani M, Invernizzi F, De Souza CF-M, Farina L, Carrara F, Lamantea E et al (2007) Novel mutations of ND genes in complex I deficiency associated with mitochondrial encephalopathy. Brain: J Neurol 130(Pt 7):1894–1904. doi:10.1093/brain/awm114

Mallilankaraman K, Doonan P, Cárdenas C, Chandramoorthy HC, Müller M, Miller R, Hoffman NE et al (2012) MICU1 is an essential gatekeeper for MCU-mediated mitochondrial Ca(2+) uptake that regulates cell survival. Cell 151(3):630–644. doi:10.1016/j.cell.2012.10.011

Manczak M, Calkins MJ, Reddy PH (2011) Impaired mitochondrial dynamics and abnormal interaction of amyloid beta with mitochondrial protein Drp1 in neurons from patients with Alzheimer's disease: implications for neuronal damage. Hum Mol Genet 20(13):2495–2509. doi:10.1093/hmg/ddr139

Martinou JC, Desagher S, Antonsson B (2000) Cytochrome c release from mitochondria: all or nothing. Nat Cell Biol 2(3):E41–E43. doi:10.1038/35004069

Matthews PM, Tampieri D, Berkovic SF, Andermann F, Silver K, Chityat D, Arnold DL (1991) Magnetic resonance imaging shows specific abnormalities in the MELAS syndrome. Neurology 41(7):1043–1046. Retrieved from http://www.ncbi.nlm.nih.gov/pubmed/2067632

Mattman A, Sirrs S, Mezei M, Salvarinova-Zivkovic R, Alfadhel M, Lillquist Y (2011) Mitochondrial disease clinical manifestations: an overview. BCMJ 53(4):183–187. Retrieved from http://www.bcmj.org/articles/mitochondrial-disease-clinical-manifestations-overview

Mayer WJ, Remy M, Rudolph G (2011) Kearns-Sayre syndrome: a mitochondrial disease (OMIM #530000). Der Ophthalmologe: Zeitschrift der Deutschen Ophthalmologischen Gesellschaft 108(5):459–462. doi:10.1007/s00347-010-2296-3

McBride HM, Neuspiel M, Wasiak S (2006) Mitochondria: more than just a powerhouse. CB Curr Biol 16(14):R551–R560. doi:10.1016/j.cub.2006.06.054

McCoy B, Owens C, Howley R, Ryan S, King M, Farrell MA, Lynch BJ (2011) Partial status epilepticus—rapid genetic diagnosis of Alpers' disease. Eur J Paediatr Neurol: EJPN: Official J Eur Paediatr Neurol Soc 15(6):558–562. doi:10.1016/j.ejpn.2011.05.012

McFarland R, Turnbull DM (2009) Batteries not included: diagnosis and management of mitochondrial disease. J Intern Med 265(2):210–228. doi:10.1111/j.1365-2796.2008.02066.x

McFarland, Robert, Taylor RW, Turnbull DM (2002) The neurology of mitochondrial DNA disease. Lancet Neurol 1(6):343–351. Retrieved from http://www.ncbi.nlm.nih.gov/pubmed/ 12849395

McKelvie PA, Morley JB, Byrne E, Marzuki S (1991) Mitochondrial encephalomyopathies: a correlation between neuropathological findings and defects in mitochondrial DNA. J Neurol Sci 102(1):51–60. Retrieved from http://www.ncbi.nlm.nih.gov/pubmed/1906931

McShane MA, Hammans SR, Sweeney M, Holt IJ, Beattie TJ, Brett EM, Harding AE (1991) Pearson syndrome and mitochondrial encephalomyopathy in a patient with a deletion of mtDNA. Am J Hum Genet 48(1):39–42. Retrieved from http://www.pubmedcentral.nih.gov/ articlerender.fcgi?artid=1682744&tool=pmcentrez&rendertype=abstract

Medina L, Chi TL, DeVivo DC, Hilal SK (1990) MR findings in patients with subacute necrotizing encephalomyelopathy (Leigh syndrome): correlation with biochemical defect. AJNR Am J Neuroradiol 11(2):379–384. Retrieved from http://www.ncbi.nlm.nih.gov/ pubmed/2156413

Misgeld T, Kerschensteiner M, Bareyre FM, Burgess RW, Lichtman JW (2007) Imaging axonal transport of mitochondria in vivo. Nat Methods 4(7):559–561. doi:10.1038/nmeth1055

Misko AL, Sasaki Y, Tuck E, Milbrandt J, Baloh RH (2012) Mitofusin2 mutations disrupt axonal mitochondrial positioning and promote axon degeneration. J Neurosci: Official J Soc Neurosci 32(12):4145–4155. doi:10.1523/JNEUROSCI.6338-11.2012

Miyata H, Silverman HS, Sollott SJ, Lakatta EG, Stern MD, Hansford RG (1991) Measurement of mitochondrial free Ca2+ concentration in living single rat cardiac myocytes. Am J Physiol 261(4 Pt 2):H1123–H1134. Retrieved from http://www.ncbi.nlm.nih.gov/pubmed/1928394

Mizushima N (2007) Autophagy: process and function. Genes Dev 21(22):2861–2873. doi:10. 1101/gad.1592207

Mizushima N, Yoshimori T, Ohsumi Y (2011) The role of Atg proteins in autophagosome formation. Annu Rev Cell Dev Biol 27:107–132. doi:10.1146/annurev-cellbio-092910-154005

Molina AJ, Shirihai OS (2009) Monitoring mitochondrial dynamics with photoactivatable [corrected] green fluorescent protein. Methods Enzymol 457:289–304. doi:10.1016/ S0076-6879(09)05016-2

Morava E, Van den Heuvel L, Hol F, De Vries MC, Hogeveen M, Rodenburg RJ, Smeitink JAM (2006) Mitochondrial disease criteria: diagnostic applications in children. Neurology 67(10):1823–1826. doi:10.1212/01.wnl.0000244435.27645.54

Moroni I, Bugiani M, Bizzi A, Castelli G, Lamantea E, Uziel G (2002) Cerebral white matter involvement in children with mitochondrial encephalopathies. Neuropediatrics 33(2):79–85. doi:10.1055/s-2002-32372

Muñoz A, Mateos F, Simón R, García-Silva MT, Cabello S, Arenas J (1999) Mitochondrial diseases in children: neuroradiological and clinical features in 17 patients. Neuroradiology 41(12):920–928. Retrieved from http://www.ncbi.nlm.nih.gov/pubmed/10639669

Nagai T, Sawano A, Park ES, Miyawaki A (2001) Circularly permuted green fluorescent proteins engineered to sense Ca2+. Proc Natl Acad Sci USA 98(6):3197–3202. doi:10.1073/pnas. 051636098

Nakada K, Inoue K, Ono T, Isobe K, Ogura A, Goto YI, Nonaka I et al (2001) Inter-mitochondrial complementation: mitochondria-specific system preventing mice from expression of disease phenotypes by mutant mtDNA. Nat Med 7(8):934–940. doi:10.1038/90976

Naviaux RK, Nguyen KV (2004) POLG mutations associated with Alpers' syndrome and mitochondrial DNA depletion. Ann Neurol 55(5):706–712. doi:10.1002/ana.20079

Nishino I (1999) Thymidine phosphorylase gene mutations in MNGIE, a human mitochondrial disorder. Science 283(5402):689–692. doi:10.1126/science.283.5402.689

Nishino I, Spinazzola A, Papadimitriou A, Hammans S, Steiner I, Hahn CD, Connolly AM et al (2000) Mitochondrial neurogastrointestinal encephalomyopathy: an autosomal recessive disorder due to thymidine phosphorylase mutations. Ann Neurol 47(6):792–800. Retrieved from http://www.ncbi.nlm.nih.gov/pubmed/10852545

Nixon RA, Wegiel J, Kumar A, Yu WH, Peterhoff C, Cataldo A, Cuervo AM (2005) Extensive involvement of autophagy in Alzheimer disease: an immuno-electron microscopy study. J Neuropathol Exp Neurol 64(2):113–122. Retrieved from http://www.ncbi.nlm.nih.gov/pubmed/15751225

O'Reilly CM, Fogarty KE, Drummond RM, Tuft RA, Walsh JV (2003) Quantitative analysis of spontaneous mitochondrial depolarizations. Biophys J 85(5):3350–3357. doi:10.1016/S0006-3495(03)74754-7

Ong S-B, Gustafsson AB (2012) New roles for mitochondria in cell death in the reperfused myocardium. Cardiovasc Res 94(2):190–196. doi:10.1093/cvr/cvr312

Ong S-B, Hausenloy DJ (2010) Mitochondrial morphology and cardiovascular disease. Cardiovasc Res 88(1):16–29. doi:10.1093/cvr/cvq237

Ong S-B, Subrayan S, Lim SY, Yellon DM, Davidson SM, Hausenloy DJ (2010) Inhibiting mitochondrial fission protects the heart against ischemia/reperfusion injury. Circulation 121(18):2012–2022. doi:10.1161/CIRCULATIONAHA.109.906610

Ong S-B, Hall AR, Hausenloy DJ (2012) Mitochondrial dynamics in cardiovascular health and disease. Antioxid Redox Signal. doi:10.1089/ars.2012.4777

Oppenheim C, Galanaud D, Samson Y, Sahel M, Dormont D, Wechsler B, Marsault C (2000) Can diffusion weighted magnetic resonance imaging help differentiate stroke from stroke-like events in MELAS? J Neurol, Neurosurg, Psychiatry 69(2):248–50. Retrieved from http://www.pubmedcentral.nih.gov/articlerender.fcgi?artid=1737057&tool=pmcentrez&rendertype=abstract

Ost A, Svensson K, Ruishalme I, Brännmark C, Franck N, Krook H, Sandström P et al (2010) Attenuated mTOR signaling and enhanced autophagy in adipocytes from obese patients with type 2 diabetes. Mol Med (Cambridge, MA) 16(7–8):235–246. doi:10.2119/molmed.2010.00023

Pauli W, Zarzycki A, Krzyształowski A, Walecka A (2013) CT and MRI imaging of the brain in MELAS syndrome. Pol J Radiol/Pol Med Soc Radiol 78(3):61–65. doi:10.12659/PJR.884010

Pavlakis SG, Phillips PC, DiMauro S, De Vivo DC, Rowland LP (1984) Mitochondrial myopathy, encephalopathy, lactic acidosis, and strokelike episodes: a distinctive clinical syndrome. Ann Neurol 16(4):481–488. doi:10.1002/ana.410160409

Pearson HA, Lobel JS, Kocoshis SA, Naiman JL, Windmiller J, Lammi AT, Hoffman R et al (1979) A new syndrome of refractory sideroblastic anemia with vacuolization of marrow precursors and exocrine pancreatic dysfunction. J Pediatr 95(6):976–984. Retrieved from http://www.ncbi.nlm.nih.gov/pubmed/501502

Perez-Atayde AR (2013) Diagnosis of mitochondrial neurogastrointestinal encephalopathy disease in gastrointestinal biopsies. Hum Pathol. Retrieved from http://www.sciencedirect.com/science/article/pii/S0046817713000026

Petty RK, Harding AE, Morgan-Hughes JA (1986) The clinical features of mitochondrial myopathy. Brain: J Neurol 109(Pt 5):915–938. Retrieved from http://www.ncbi.nlm.nih.gov/pubmed/3779373

Phadwal K, Alegre-Abarrategui J, Watson AS, Pike L, Anbalagan S, Hammond EM, Wade-Martins R et al (2012) A novel method for autophagy detection in primary cells: impaired levels of macroautophagy in immunosenescent T cells. Autophagy 8(4):677–689. doi:10.4161/auto.18935

Pinton P, Rizzuto R (2006) Bcl-2 and Ca2+ homeostasis in the endoplasmic reticulum. Cell Death Differ 13(8):1409–1418. doi:10.1038/sj.cdd.4401960

Plovanich M, Bogorad RL, Sancak Y, Kamer KJ, Strittmatter L, Li AA, Girgis HS et al (2013) MICU2, a paralog of MICU1, resides within the mitochondrial uniporter complex to regulate calcium handling. PLoS ONE 8(2):e55785. doi:10.1371/journal.pone.0055785

Plucińska G, Paquet D, Hruscha A, Godinho L, Haass C, Schmid B, Misgeld T (2012) In vivo imaging of disease-related mitochondrial dynamics in a vertebrate model system. J Neurosci: Official J Soc Neurosci 32(46):16203–16212. doi:10.1523/JNEUROSCI.1327-12.2012

Putney JW, Thomas AP (2006) Calcium signaling: double duty for calcium at the mitochondrial uniporter. CB Curr Biol 16(18):R812–R815. doi:10.1016/j.cub.2006.08.040

Quinonez SC, Leber SM, Martin DM, Thoene JG, Bedoyan JK (2013) Leigh syndrome in a girl with a novel DLD mutation causing E3 deficiency. Pediatr Neurol. Retrieved from http://www.sciencedirect.com/science/article/pii/S088789941200450X

Rahman S, Blok RB, Dahl HH, Danks DM, Kirby DM, Chow CW, Christodoulou J et al (1996) Leigh syndrome: clinical features and biochemical and DNA abnormalities. Ann Neurol 39(3):343–351. doi:10.1002/ana.410390311

Rahman S, Brown RM, Chong WK, Wilson CJ, Brown GK (2001) A SURF1 gene mutation presenting as isolated leukodystrophy. Ann Neurol 49(6):797–800. Retrieved from http://www.ncbi.nlm.nih.gov/pubmed/11409433

Rahn JJ, Stackley KD, Chan SSL (2013) Opa1 is required for proper mitochondrial metabolism in early development. PLoS ONE 8(3):e59218. doi:10.1371/journal.pone.0059218

Ramonet D, Perier C, Recasens A, Dehay B, Bové J, Costa V, Scorrano L et al (2013) Optic atrophy 1 mediates mitochondria remodeling and dopaminergic neurodegeneration linked to complex I deficiency. Cell Death Differ 20(1):77–85. doi:10.1038/cdd.2012.95

Rand SD, Prost R, Li SJ (1999) Proton MR spectroscopy of the brain. Neuroimaging Clin North Am 9(2):379–395. Retrieved from http://www.ncbi.nlm.nih.gov/pubmed/10318721

Renard D, Campello C, Le Floch A, Castelnovo G, Taieb G (2012) Globus pallidus and substantia nigra hypointensities on T2-weighted imaging in MELAS. J Neurol 259(12):2720–2722. doi:10.1007/s00415-012-6633-0

Rizzuto Rosario, Pozzan T (2006) Microdomains of intracellular Ca2+: molecular determinants and functional consequences. Physiol Rev 86(1):369–408. doi:10.1152/physrev.0 0004.2005

Rizzuto R, Brini M, Murgia M, Pozzan T (1993) Microdomains with high Ca2+ close to IP3-sensitive channels that are sensed by neighboring mitochondria. Science (New York, NY) 262(5134):744–747. Retrieved from http://www.ncbi.nlm.nih.gov/pubmed/8235595

Rodriguez-Enriquez S, He L, Lemasters JJ (2004) Role of mitochondrial permeability transition pores in mitochondrial autophagy. Int J Biochem Cell Biol 36(12):2463–2472. doi:10.1016/j.biocel.2004.04.009

Rodriguez-Enriquez S, Kim I, Currin RT, Lemasters JJ (2006) Tracker dyes to probe mitochondrial autophagy (mitophagy) in rat hepatocytes. Autophagy 2(1):39–46. Retrieved from http://www.ncbi.nlm.nih.gov/pubmed/16874071

Rosado CJ, Mijaljica D, Hatzinisiriou I, Prescott M, Devenish RJ (2008) Rosella: a fluorescent pH-biosensor for reporting vacuolar turnover of cytosol and organelles in yeast. Autophagy 4(2):205–213. Retrieved from http://www.ncbi.nlm.nih.gov/pubmed/18094608

Ross BD (2000) Real or imaginary? Human metabolism through nuclear magnetism. IUBMB Life 50(3):177–187. doi:10.1080/152165400300001499

Rossmanith W, Freilinger M, Roka J, Raffelsberger T, Moser-Thier K, Prayer D, Bernert G et al (2008) Isolated cytochrome c oxidase deficiency as a cause of MELAS. J Med Genet 45(2):117–121. doi:10.1136/jmg.2007.052076

Rotig A, Colonna M, Bonnefont JP, Blanche S, Fischer A, Saudubray JM, Munnich A (1989) Mitochondrial DNA deletion in Pearson's marrow/pancreas syndrome. Lancet 1(8643):902–903. Retrieved from http://www.ncbi.nlm.nih.gov/pubmed/2564980

Rötig A, Lebon S, Zinovieva E, Mollet J, Sarzi E, Bonnefont J-P, Munnich A (2004) Molecular diagnostics of mitochondrial disorders. Biochim Biophys Acta 1659(2–3):129–135. doi:10.1016/j.bbabio.2004.07.007

Rouzier C, Bannwarth S, Chaussenot A, Chevrollier A, Verschueren A, Bonello-Palot N, Fragaki K et al (2012) The MFN2 gene is responsible for mitochondrial DNA instability and optic atrophy "plus" phenotype. Brain: J Neurol 135(Pt 1):23–34. doi:10.1093/brain/awr323

Rudnicki DD, Pletnikova O, Vonsattel J-PG, Ross CA, Margolis RL (2008) A comparison of huntington disease and huntington disease-like 2 neuropathology. J Neuropathol Exp Neurol 67(4):366–374. doi:10.1097/NEN.0b013e31816b4aee

Rutter J, Winge DR, Schiffman JD (2010) Succinate dehydrogenase—Assembly, regulation and role in human disease. Mitochondrion 10(4):393–401. Retrieved from http://www.sciencedirect.com/science/article/pii/S1567724910000358

Saitoh S, Momoi MY, Yamagata T, Mori Y, Imai M (1998) Effects of dichloroacetate in three patients with MELAS. Neurology 50(2):531–534. Retrieved from http://www.ncbi.nlm.nih.gov/pubmed/9484392

Saks V, Favier R, Guzun R, Schlattner U, Wallimann T (2006) Molecular system bioenergetics: regulation of substrate supply in response to heart energy demands. J Physiol 577(Pt 3):769–777. doi:10.1113/jphysiol.2006.120584

Saneto RP, Friedman SD, Shaw DWW (2008) Neuroimaging of mitochondrial disease. Mitochondrion 8(5–6):396–413. doi:10.1016/j.mito.2008.05.003

Santel A, Frank S, Gaume B, Herrler M, Youle RJ, Fuller MT (2003) Mitofusin-1 protein is a generally expressed mediator of mitochondrial fusion in mammalian cells. J Cell Sci 116(Pt 13):2763–2774. doi:10.1242/jcs.00479

Scaglia F, Wong L-JC, Vladutiu GD, Hunter JV (2005) Predominant cerebellar volume loss as a neuroradiologic feature of pediatric respiratory chain defects. AJNR Am J Neuroradiol 26(7):1675–1680. Retrieved from http://www.ncbi.nlm.nih.gov/pubmed/16091512

Schapira AHV (2006) Mitochondrial disease. Lancet 368(9529):70–82. doi:10.1016/S0140-6736(06)68970-8

Schauss AC, Huang H, Choi S-Y, Xu L, Soubeyrand S, Bilodeau P, Zunino R et al (2010) A novel cell-free mitochondrial fusion assay amenable for high-throughput screenings of fusion modulators. BMC Biol 8:100. doi:10.1186/1741-7007-8-100

Schreur JH, Figueredo VM, Miyamae M, Shames DM, Baker AJ, Camacho SA (1996) Cytosolic and mitochondrial [Ca2+] in whole hearts using indo-1 acetoxymethyl ester: effects of high extracellular Ca2+. Biophys J 70(6):2571–2580. doi:10.1016/S0006-3495(96)79828-4

Schuelke M, Smeitink J, Mariman E, Loeffen J, Plecko B, Trijbels F, Stöckler-Ipsiroglu S et al (1999) Mutant NDUFV1 subunit of mitochondrial complex I causes leukodystrophy and myoclonic epilepsy. Nat Genet 21(3):260–261. doi:10.1038/6772

Sedova M, Dedkova EN, Blatter LA (2006) Integration of rapid cytosolic Ca2+ signals by mitochondria in cat ventricular myocytes. Am J Physiol Cell Physiol 291(5):C840–C850. doi:10.1152/ajpcell.0 0619.2005

Sekiya S, Tanaka M, Hayashi S, Oyanagi S (1982) Light- and electron-microscopic studies of intracytoplasmic acidophilic granules in the human locus coeruleus and substantia nigra. Acta Neuropathol 56(1):78–80. Retrieved from http://www.ncbi.nlm.nih.gov/pubmed/6278812

Shaibani A, Shchelochkov OA, Zhang S, Katsonis P, Lichtarge O, Wong L-J, Shinawi M (2009) Mitochondrial neurogastrointestinal encephalopathy due to mutations in RRM2B. Arch Neurol 66(8):1028–1032. doi:10.1001/archneurol.2009.139

Shoubridge EA (2001) Nuclear genetic defects of oxidative phosphorylation. Hum Mol Genet 10(20):2277–2284. Retrieved from http://www.ncbi.nlm.nih.gov/pubmed/11673411

Sidiropoulos C, Moro E, Lang AE (2013) Extensive intracranial calcifications in a patient with a novel polymerase γ-1 mutation. Neurology 81(2):197–198. doi:10.1212/WNL.0b013e31829a3438

Sigmond T, Fehér J, Baksa A, Pásti G, Pálfia Z, Takács-Vellai K, Kovács J et al (2008) Qualitative and quantitative characterization of autophagy in Caenorhabditis elegans by electron microscopy. Methods Enzymol 451:467–491. doi:10.1016/S0076-6879(08)03228-X

Sijens PE, Smit GPA, Rödiger LA, Van Spronsen FJ, Oudkerk M, Rodenburg RJ, Lunsing RJ (2008) MR spectroscopy of the brain in Leigh syndrome. Brain Dev 30(9):579–583. doi:10.1016/j.braindev.2008.01.011

Slodzinski MK, Aon MA, O'Rourke B (2008) Glutathione oxidation as a trigger of mitochondrial depolarization and oscillation in intact hearts. J Mol Cell Cardiol 45(5):650–660. doi:10.1016/j.yjmcc.2008.07.017

Smirnova E, Shurland DL, Ryazantsev SN, Van der Bliek AM (1998) A human dynamin-related protein controls the distribution of mitochondria. J Cell Biol 143(2):351–358. Retrieved from http://www.pubmedcentral.nih.gov/articlerender.fcgi?artid=2132828&tool=pmcentrez&rendertype=abstract

Sproule DM, Kaufmann P (2008) Mitochondrial encephalopathy, lactic acidosis, and strokelike episodes: basic concepts, clinical phenotype, and therapeutic management of MELAS syndrome. Ann N Y Acad Sci 1142:133–158. doi:10.1196/annals.1444.011

Stanley WC, Recchia FA, Lopaschuk GD (2005) Myocardial substrate metabolism in the normal and failing heart. Physiol Rev 85(3):1093–1129. doi:10.1152/physrev.00006.2004

Sun F, Huo X, Zhai Y, Wang A, Xu J, Su D, Bartlam M et al (2005) Crystal structure of mitochondrial respiratory membrane protein complex II. Cell 121(7):1043–1057. doi:10.1016/j.cell.2005.05.025

Tam EWY, Feigenbaum A, Addis JBL, Blaser S, Mackay N, Al-Dosary M, Taylor RW et al (2008) A novel mitochondrial DNA mutation in COX1 leads to strokes, seizures, and lactic acidosis. Neuropediatrics 39(6):328–334. doi:10.1055/s-0029-1202287

Tang S, Wang J, Lee N-C, Milone M, Halberg MC, Schmitt ES, Craigen WJ et al (2011) Mitochondrial DNA polymerase gamma mutations: an ever expanding molecular and clinical spectrum. J Med Genet 48(10):669–681. doi:10.1136/jmedgenet-2011-100222

Tang S, Dimberg EL, Milone M, Wong L-JC (2012) Mitochondrial neurogastrointestinal encephalomyopathy (MNGIE)-like phenotype: an expanded clinical spectrum of POLG1 mutations. J Neurol 259(5):862–868. doi:10.1007/s00415-011-6268-6

Tanigawa J, Kaneko K, Honda M, Harashima H, Murayama K, Wada T, Takano K et al (2012) Two Japanese patients with Leigh syndrome caused by novel SURF1 mutations. Brain Dev. Retrieved from http://www.sciencedirect.com/science/article/pii/S0387760412000393

Tanji K, DiMauro S, Bonilla E (1999) Disconnection of cerebellar Purkinje cells in Kearns-Sayre syndrome. J Neurol Sci 166(1):64–70. Retrieved from http://www.ncbi.nlm.nih.gov/pubmed/10465502

Thorburn DR, Sugiana C, Salemi R, Kirby DM, Worgan L, Ohtake A, Ryan MT (2004) Biochemical and molecular diagnosis of mitochondrial respiratory chain disorders. Biochim Biophys Acta 1659(2–3):121–128. doi:10.1016/j.bbabio.2004.08.006

Tian X, Ma X, Qiao D, Ma A, Yan F, Huang X (2005) mCICR is required for As2O3-induced permeability transition pore opening and cytochrome c release from mitochondria. Mol Cell Biochem 277(1–2):33–42. doi:10.1007/s11010-005-4818-x

Tinel H, Cancela JM, Mogami H, Gerasimenko JV, Gerasimenko OV, Tepikin AV, Petersen OH (1999) Active mitochondria surrounding the pancreatic acinar granule region prevent spreading of inositol trisphosphate-evoked local cytosolic Ca(2+) signals. EMBO J 18(18):4999–5008. doi:10.1093/emboj/18.18.4999

Trimmer PA, Swerdlow RH, Parks JK, Keeney P, Bennett JP, Miller SW, Davis RE et al (2000) Abnormal mitochondrial morphology in sporadic Parkinson's and Alzheimer's disease cybrid cell lines. Exp Neurol 162(1):37–50. doi:10.1006/exnr 2000.7333

Twig G, Graf SA, Wikstrom JD, Mohamed H, Haigh SE, Elorza A, Deutsch M et al (2006) Tagging and tracking individual networks within a complex mitochondrial web with photoactivatable GFP. Am J Physiol Cell Physiol 291(1):C176–C184. doi:10.1152/ajpcell.00348.2005

Twig G, Elorza A, Molina AJA, Mohamed H, Wikstrom JD, Walzer G, Stiles L et al (2008) Fission and selective fusion govern mitochondrial segregation and elimination by autophagy. EMBO J 27(2):433–446. doi:10.1038/sj.emboj.7601963

Tzoulis C, Engelsen BA, Telstad W, Aasly J, Zeviani M, Winterthun S, Ferrari G et al (2006) The spectrum of clinical disease caused by the A467T and W748S POLG mutations: a study of 26 cases. Brain: J Neurol 129(Pt 7):1685–1692. doi:10.1093/brain/awl097

Uusimaa J, Gowda V, McShane A, Smith C, Evans J, Shrier A, Narasimhan M et al (2013) Prospective study of POLG mutations presenting in children with intractable epilepsy: prevalence and clinical features. Epilepsia 54(6):1002–1011. doi:10.1111/epi.12115

Valanne L, Ketonen L, Majander A, Suomalainen A, Pihko H (1998) Neuroradiologic findings in children with mitochondrial disorders. AJNR Am J Neuroradiol 19(2):369–377. Retrieved from http://www.ncbi.nlm.nih.gov/pubmed/9504497

Van Goethem G, Luoma P, Rantamäki M, Al Memar A, Kaakkola S, Hackman P, Krahe R et al (2004) POLG mutations in neurodegenerative disorders with ataxia but no muscle involvement. Neurology 63(7):1251–1257. Retrieved from http://www.ncbi.nlm.nih.gov/pubmed/15477547

Wang X, Su B, Lee H, Li X, Perry G, Smith MA, Zhu X (2009) Impaired balance of mitochondrial fission and fusion in Alzheimer's disease. J Neurosci: Official J Soc Neurosci 29(28):9090–9103. doi:10.1523/JNEUROSCI.1357-09.2009

Wilichowski E, Pouwels PJ, Frahm J, Hanefeld F (1999) Quantitative proton magnetic resonance spectroscopy of cerebral metabolic disturbances in patients with MELAS. Neuropediatrics 30(5):256–263. doi:10.1055/s-2007-973500

Williams GSB, Boyman L, Chikando AC, Khairallah RJ, Lederer WJ (2013) Mitochondrial calcium uptake. Proc Natl Acad Sci USA 110(26):10479–10486. doi:10.1073/pnas.1300410110

Wray SH, Provenzale JM, Johns DR, Thulborn KR (1995) MR of the brain in mitochondrial myopathy. AJNR Am J Neuroradiol 16(5):1167–1173. Retrieved from http://www.ncbi.nlm.nih.gov/pubmed/7639148

Wu J-Q, Pollard TD (2005) Counting cytokinesis proteins globally and locally in fission yeast. Science (New York, NY) 310(5746):310–314. doi:10.1126/science.1113230

Yablonskiy DA, Neil JJ, Raichle ME, Ackerman JJH (1998) Homonuclear J coupling effects in volume localized NMR spectroscopy: pitfalls and solutions. Magn Reson Med 39(2):169–178. doi:10.1002/mrm.1910390202

Yilmaz A, Gdynia H-J, Ponfick M, Rösch S, Lindner A, Ludolph AC, Sechtem U (2012) Cardiovascular magnetic resonance imaging (CMR) reveals characteristic pattern of myocardial damage in patients with mitochondrial myopathy. Clinical Res Cardiol: Official J German Cardiac Soc 101(4):255–261. doi:10.1007/s00392-011-0387-z

Ylä-Anttila P, Vihinen H, Jokitalo E, Eskelinen E-L (2009a) 3D tomography reveals connections between the phagophore and endoplasmic reticulum. Autophagy 5(8):1180–1185. Retrieved from http://www.ncbi.nlm.nih.gov/pubmed/19855179

Ylä-Anttila P, Vihinen H, Jokitalo E, Eskelinen E-L (2009b) Monitoring autophagy by electron microscopy in Mammalian cells. Methods Enzymol 452:143–164. doi:10.1016/S0076-6879(08)03610-0

Yonemura K, Hasegawa Y, Kimura K, Minematsu K, Yamaguchi T (2001) Diffusion-weighted MR imaging in a case of mitochondrial myopathy, encephalopathy, lactic acidosis, and strokelike episodes. AJNR Am J Neuroradiol 22(2):269–272. Retrieved from http://www.ncbi.nlm.nih.gov/pubmed/11156767

Zeviani Massimo, Di Donato S (2004) Mitochondrial disorders. Brain: J Neurol 127(Pt 10):2153–2172. doi:10.1093/brain/awh259

Zeviani M, Moraes CT, DiMauro S, Nakase H, Bonilla E, Schon EA, Rowland LP (1988) Deletions of mitochondrial DNA in Kearns-Sayre syndrome. Neurology 38(9):1339–1346. Retrieved from http://www.ncbi.nlm.nih.gov/pubmed/3412580

Zhou C, Zhong W, Zhou J, Sheng F, Fang Z, Wei Y, Chen Y et al (2012) Monitoring autophagic flux by an improved tandem fluorescent-tagged LC3 (mTagRFP-mWasabi-LC3) reveals that high-dose rapamycin impairs autophagic flux in cancer cells. Autophagy 8(8):1215–1226. doi:10.4161/auto.20284

Zhu Z, Yao J, Johns T, Fu K, De Bie I, Macmillan C, Cuthbert AP et al (1998) SURF1, encoding a factor involved in the biogenesis of cytochrome c oxidase, is mutated in Leigh syndrome. Nat Genet 20(4):337–343. doi:10.1038/3804

Züchner S, Mersiyanova IV, Muglia M, Bissar-Tadmouri N, Rochelle J, Dadali EL, Zappia M et al (2004) Mutations in the mitochondrial GTPase mitofusin 2 cause Charcot-Marie-Tooth neuropathy type 2A. Nat Genet 36(5):449–451. doi:10.1038/ng1341

Chapter 6
A Novel Hybrid Magnetoacoustic Measurement Method for Breast Cancer Detection

Maheza Irna Mohamad Salim, Nugraha Priya Utama, Eko Supriyanto, Khin Wee Lai, Yan Chai Hum and Yin Mon Myint

Abstract Breast cancer is the most common cancer in women worldwide. It is a disease of uncontrolled breast cells growth, in which the cells acquire genetic alteration, causing them to proliferate more aggressively as compared to normal tissue development. In current medical practice, the gold standard for breast cancer screening is mammography. However, its usage is unsafe as it exposes patient to ionizing radiation and it is less comfortable due to the need for breast compression. Another available option for breast screening is ultrasound. To date, ultrasound is an important adjunct modality to mammography regardless of its low sensitivity in detecting small cancers from normal tissues due to overlapping ultrasonic characteristics of these tissues. To address this problem, a hybrid magnetoacoustic measurement method (HMM) that combines ultrasound and magnetism for the

M. I. Mohamad Salim (✉)
Universiti Teknologi Malaysia, Level 3, V01, Block A, Satellite Building, Skudai-Johor, Malaysia
e-mail: maheza@biomedical.utm.my

N. P. Utama · Y. M. Myint
Faculty of Biosciences and Medical Engineering, Universiti Teknologi Malaysia, Level 3, V01, Block A, Satellite Building, Skudai-Johor, Malaysia
e-mail: utama@biomedical.utm.my

Y. M. Myint
e-mail: yinmontt@gmail.com

E. Supriyanto
Faculty of Biosciences and Medical Engineering, IJN-UTM Cardiovascular Engineering Center, Universiti Teknologi Malaysia, Skudai-Johor, Malaysia
e-mail: eko@biomedical.utm.my

K. W. Lai
Biomedical Engineering Department, University Malaya, Kuala Lumpur, Malaysia
e-mail: lai.khinwee@um.edu.my

Y. C. Hum
MIMOS Berhad, Technology Park Malaysia, 57000 Kuala Lumpur, Malaysia
e-mail: yc.hum@mimos.my

K. W. Lai et al., *Advances in Medical Diagnostic Technology*,
Lecture Notes in Bioengineering, DOI: 10.1007/978-981-4585-72-9_6,
© Springer Science+Business Media Singapore 2014

simultaneous assessment of bioelectric and acoustic profiles of breast tissue is proposed. Previous studies have shown that in cancerous tissue, changes in ultrasonic characteristics occur due to uncontrolled cell multiplication, excessive accumulation of protein in stroma, and enhancement of capillary density. Additionally, changes in conductivity also occur due to the increase in cellular water and electrolyte content, as well as membrane permeability due to increased metabolic requirements. In HMM, the interaction between the ultrasound wave and the magnetic field in the breast tissue results in Lorentz force. This produces a magnetoacoustic voltage output, which is proportional to breast tissue conductivity. Simultaneously, the ultrasound wave is sensed back by the ultrasound receiver for tissue acoustic evaluation. At the end of this study, ultrasound wave characterization results showed that normal breast tissue experienced higher attenuation compared with cancerous tissue. The mean magnetoacoustic voltage results for normal tissue were lower than the cancerous tissue group. This demonstrates that the combination of acoustic and bioelectric measurements appears to be a promising approach for diagnosis.

6.1 Introduction and Literature Review

6.1.1 Breast Cancer

Breast cancer is a disease of uncontrolled breast cells growth, in which the cells acquire a genetic alteration that allows them to multiply and grow outside the context of normal tissue development (Locasale and Cantley 2010). The cell metabolism increases to meet the requirements of rapid cell proliferation, autonomous cell growth, and cell survival (Locasale and Cantley 2010; Alberts et al. 2002).

The key aspect in diagnosis of breast cancer is to determine whether the cancer is in situ or invasive (Donegan and Spratt 2002; Moinfar 2007; Cuzick 2003). In situ cancers confine themselves to the ducts or lobules and do not spread to the surrounding organ. Two main types of in situ breast cancer are the ductal carcinoma in situ (DCIS) and lobular carcinoma in situ (LCIS). DCIS means that the abnormal cancer cells are found only at the lining of the ducts. However, it can be found in more than one place in the breast since the cancer travels through the ducts. DCIS has a high cure rate especially if given early treatments. However, it can change to invasive carcinoma without a proper treatment. LCIS means that the abnormal cancer cells are found in the lining of milk lobules and is a warning sign of increased risk of developing invasive cancers. Invasive cancer is a cancer that has penetrated through normal tissue barriers and invades the surrounding organs via the bloodstream and the lymphatic system. The most common invasive cancers are invasive lobular carcinoma and invasive ductal carcinoma. There are also some rare cancers such as the inflammatory breast cancer and Paget's disease that differ from invasive ductal and lobular carcinoma, in which they do not form a distinct mass or lump in the breast (Donegan and Spratt 2002; Moinfar 2007).

The most common symptom of breast cancer is the presence of painless and slow growing lump that may alter the contour or size of the breast (Thomas et al. 1997). It is also characterized by skin changes, inverted nipple, and bloodstained nipple discharge (Donegan and Spratt 2002; Moinfar 2007). The lymphatic nodes under the armpit may be swollen if affected by cancer. In late stages, the growth may ulcerate through the skin and get infected (Donegan and Spratt 2002; Moinfar 2007). Bone pain, tenderness over the liver, headaches, shortness of breath, and chronic cough may be an indication of the cancer spreading to the other organs in the body (Moinfar 2007).

The main risk factor for breast cancer can be usefully grouped into four major categories (Cuzick 2003): family history or genetics, hormonal, proliferative breast benign pathology, and mammographic density. These four factors have now been thoroughly studied, and accurate quantitative estimates are available for the factors (Cuzick 2003). In terms of genetics, the mutations of BRCA1 and BRCA2 genes have been identified as genetic susceptibility of breast cancer in which carriers of the genes have at least 40–85 % chances of getting breast cancer (Cuzick 2003). Besides that, several lines of evident points to estrogen levels as a hormonal prime factor for the development of breast cancer (Cuzick 2003). The evidents include result of laboratory studies, direct measurement to postmenopausal women, and risk reduction when women take anti-estrogen (Thomas et al. 1997). However, details of mechanisms are still unclear. In addition to that, the risk of cancer following benign breast disease has also been identified. A recent study shows that benign breast disease in the absence of proliferation does not carry any excess risks (Cuzick 2003). However, a simple hyperplasia doubles the excess rate, and atypical hyperplasia increases the risk of getting breast cancer to fourfold (Cuzick 2003). In terms of mammographic density, earlier studies had clearly demonstrated that a radiographically opaque area in the mammography is an important measure of the risk of developing breast cancer.

Finally, female breast has a special place in human affairs beyond its biological function. It was a prominent feature of motherhood, beauty, fertility, and abundance since the early days. Diseases of the breast particularly cancer are not only the threats to women's health and well-being but are also attacks on femininity, nurturance, motherhood, and personal identity. Hence, efforts to improve breast cancer detection and treatment must continue not only to save lives but also as a part of the social betterment.

6.1.2 Normal and Cancerous Breast Tissue: Changes in Density and Conductivity

The mammary gland is a complex tissue that consists of epithelial parenchyma embedded in an array of stromal cell (Arendt et al. 2010). It undergoes dynamic changes over the lifetime of a woman from the expanded development at puberty, to proliferation and apoptosis during the menstrual cycle and to full lobuloalveolar development for lactation (Donegan and Spratt 2002; Moinfar 2007; Riordan 2005). Previous studies had shown that changes in breast tissue density and conductivity occur in cancerous tissue. Breast carcinoma causes the breast cells to proliferate, grow, and pile up outside the context of normal tissue development, which finally results in increased local cell density (Locasale and Cantley 2010). In addition to that, it is well established that stroma associated with normal mammary gland development is totally different from that associated with carcinoma (Arendt et al. 2010; Bissel and Radisky 2001; Cukierman 2004; Orimo et al. 2005). Compared to normal breast tissue, the stroma accompanying breast carcinoma contains increased protein, immune cell infiltrates, and enhanced capillary density (Arendt et al. 2010). Extensive multiproteins accumulation in the stroma has also been associated with enhanced growth and invasiveness of the carcinoma (Shinoji et al. 1998; Provenzano et al. 2008). Increased collagen 1 and fibrin deposition, elevated expression of alpha smooth muscle actin (αSMA), collagen IV, prolyl-4-hydroxylase, fibroblast-activated protein (FAP), tenascin, desmin, calponin, caldesmon, and others have collectively altered the structure, stiffness, and density of the extracellular fluid (Arendt et al. 2010; Bissel and Radisky 2001; Cukierman 2004; Orimo et al. 2005). Enhanced capillary density or angiogenesis is the complex process, leading to the formation of new blood vessels from the preexisting vascular network and further increasing the compactness of the tissue (Gasparini 2001). The formation of angiogenesis is induced by the secretion of specific endothelial cell growth factors produced by the tumor or the stromal cells (Gasparini 2001). Studies have shown that angiogenesis plays an important role in facilitating further tumor progression (Chan et al. 2005; Shinoji et al. 1998).

In medical imaging, changes in breast density due to carcinoma are usually assessed using mammography and ultrasound. Mammographic density refers to the relative abundance of low-density adipose tissue to high-density glandular and fibroblastic stromal tissues within the breast. A previous study had shown that the involvement of 60 % or more of the breast with mammographically dense tissue confers threefold to fivefold increased risk of breast cancer (Cuzick 2003; Arendt et al. 2010). In ultrasonography, changes in tissue density are indicated by the changes in velocity. Ultrasound velocity increases when it travels through a dense material and decreases when it travels through a less dense material (Glide et al. 2007). The study report is in agreement with an earlier observation that shows ultrasound velocity traveling through breast carcinoma is higher than those of normal tissue (Bamber 1983).

The presented literature supports the fact of density alteration in breast carcinoma. In general, the density of mammary fat pad is 928 kg/m^3 and 1,020 kg/m^3 for normal tissue. However, due to the altered density of breast carcinoma, many researches (Fern 2007; Degen et al. 2007) estimate the density of breast carcinoma to be very close to muscle, which is 1,041 kg/m^3 (Fern 2007).

On the other hand, bioelectric measurement for human breast tissue has started since the 1920s with the measurement of excised normal and cancerous breast tissues. Compared to normal tissue, malignant tissue has higher conductivity

(Surowiec et al. 1988; Chaudary et al. 1984) and permittivity (Chaudary et al. 1984; Sha et al. 2002; Jossinet et al. 1985) and lower impedivity (Jossinet 1996). These changes are due to the increase in cellular water and electrolyte content as well as altered membrane permeability and blood perfusion (Zou and Guo 2003; Sha et al. 2002; Jossinet et al. 1985; Jossinet 1996). In terms of conductivity changes, a study by Chaudary et al. (1984) in the frequency range of 3 MHz–3 GHz showed that conductivity, σ, of malignant tissue is higher than that of normal tissue, particularly at frequencies below 100 MHz. The research reveals that σ is from 1.5 to 3 mS/cm for normal tissue and from 7.5 to 12 mS/cm for malignant tissue.

At the frequency of 20 kHz–100 MHz, comparative bioelectric study (Surowiec et al. 1988) between tumor and its peripheral tissue shows that cancerous tissue has higher conductivity, σ, than the surrounding tissue. Data from a few tumor samples indicate that σ ranges from 0.3 to 0.4 mS/cm for normal and from 2.0 to 8.0 for the central part of tumor. At 10 MHz specifically, normal tissue conductivity ranges between 0.3 mS/cm, while cancerous tissue conductivity ranges from 4 to 6 mS/cm.

From these measurements, it can be observed that there are significant differences in conductivity between the normal and malignant tissues.

6.1.3 Ultrasound in Breast Oncology Diagnostics

Breast ultrasound is an interactive breast imaging process using sound wave at the frequency of 20 kHz–200 MHz (Kremkau 2002). In the world of medical diagnostics, breast ultrasound has an established and significant role in the diagnostics of breast abnormalities (Svensson 1997). Ultrasound is superior from mammography for its non-ionizing radiation. This makes ultrasound an imaging of choice to manage symptomatic breast in younger women as well as in pregnant and lactating mother whom the radiation of mammography is pertinent (The and Wilson 1998). Ultrasonography is also a reliable modality for solid and cystic breast anomaly differentiation (Svensson 1997; Stavros et al. 1995; Sehgal et al. 2006). It is also used in imaging augmented and inflamed breast (Svensson 1997; Stavros 2004). However, in the current practice, the proportion of patient in whom breast ultrasonography is considered necessary is only 40 % (Flobbe et al. 2002). This means that ultrasonography is not indicated for the rest 60 % of patients referred for breast imaging (Flobbe et al. 2002). This practice explains major constraint of ultrasonography in breast imaging that limits its usage for the diagnostics of breast symptoms and for screening asymptomatic patients (The and Wilson 1998; Sehgal et al. 2006).

The major problem of ultrasonography is its low sensitivity in detecting small and preinvasive breast cancers (Fornage et al. 1990) from normal tissues due to the overlapping ultrasonic characteristics in these tissues (Bamber 1983; Edmonds and Mortensen 1991; Landini and Sarnelli 1986). Breast ultrasound diagnostics relies on several sonographic features that are based on margin, shape, and echotexture. Breast cancers are often characterized by poorly defined margins, irregular borders, spiculation, marked hyperechogenicity, shadowing, and duct extension (Stavros 2004).

A systematic review on 22 independent studies to investigate the sensitivity of ultrasound in breast cancer detection was conducted by Flobbe et al. (2002). In the review, the patient population was divided into four groups, namely (1) patient undergoes mammography. Hence, ultrasound interpretation is with the knowledge to prior mammography (five studies). (2) Patient undergoes mammography and clinical examinations. Hence, interpretation is based on the previous clinical and imaging data (four studies). (3) Patients are referred for pathology and mammography. Hence, ultrasound interpretation is with the knowledge to prior mammography and pathology result (six studies), and finally, (4) ultrasound is interpreted blindly without prior patient clinical data (seven studies). The average ultrasound sensitivity for each group of patient is 82.60, 88.25, 86.83, and 82.57 %, respectively. This systematic review has revealed the weakness of ultrasound in the diagnostics of patients with breast abnormalities regardless of the existing patients' prior clinical information. The study concludes that a little evidence support was found to confirm the well-recognized value of ultrasonography in breast cancer detection. Other than the review, an independent report by Singh et al. (2008) also shows the low sensitivity of ultrasound in the detection of breast cancer.

Another limitation of ultrasound is its inability to detect microcalcification, a calcium residue found in the breast tissue as an early indicator of DCIS (Sehgal et al. 2006; Anderson et al. 2000). In ultrasonography, the presence of microcalcification in tissue is often masked by the breast tissue heterogeneity and grainy noise due to speckle phenomena (Weinstein et al. 2002; Alizad et al. 2004). The reasons make microcalcification detection with ultrasonography unreliable (Anderson et al. 2000).

Previous study (Chang et al. 2011) also reported the sensitivity of ultrasonography for breast cancer detection evaluated by three different radiologists with experienced from 8 to 16 years. The result showed that the achieved sensitivities were 66.7, 87.5, and 56.3 % for the three radiologists. This study found that breast ultrasound diagnosis was complicated not only by the low sensitivity of the ultrasound itself but also by the dependency of ultrasound result to operator. This means that a single sonographic image may be interpreted differently by different operators and the result is relative to the operator skills and experience, variations in human perceptions of the images, differences in features used in diagnosis, and lack of quantitative measures used for image analysis (Kuo et al. 2002; Horsch et al. 2004; Wen et al. 1997, 1998; Wen and Bennet 2000; Su et al. 2007; Norton 2003; Ibrahim 2005).

This inter-reader variability has led to automated ultrasonographic image evaluation via computer-aided diagnosis (CAD system). CAD is a multistep process that involves identification of lesion by segmentation, extraction, and recognition using a complex and intelligent algorithm based on echotexture, margin, and shape (Kuo et al. 2002; Horsch et al. 2004). It offers potentially accurate judgment to generate valuable second opinion in assisting diagnosis. In CAD, the area under the ROC curve is the performance metric to evaluate CAD with one representing perfect performance. Studies have shown that sonographic

CAD is able to give a good classification performance of 0.83–0.87 (Shankar et al. 2002; Sahiner et al. 2004; Dumane et al. 2002), excellent performance of 0.92 (Sahiner et al. 2004), and near-perfect performance of 0.95–0.98 (Chen et al. 2003). With the increasing acceptance of Mammo CAD and MRI CAD, sonographic CAD has also been widely accepted to assist in diagnostics. In addition to that, previous studies also reported that sonographic CAD is helpful for diagnosis (Kuo et al. 2002; Horsch et al. 2004).

Although breast ultrasound diagnosis has improved over the time, its usage in breast cancer detection is still limited due to its low detection sensitivity to breast masses and microcalcification as well as inter-reader variability. Hence, in order for ultrasound to compete with other breast imaging modalities, additional tissue properties need to be further explored for a better breast cancer detection method.

6.1.4 Lorentz Force-Based Magnetoacoustic Imaging

The earliest research in Lorentz force-based magnetoacoustic imaging was started since 1988 by the work of Towe and Islam in the development of noninvasive measurement system for bioelectric current (Towe and Islam 1988; Islam and Towe 1988). The measurement system manipulates vibrations that are produced by Lorentz force as a result of the interaction between audible-frequency oscillating magnetic field and static magnetic field on current carrying media. In this approach, oscillating magnetic excitation is employed to induce eddy current in a conductive sample, which is put in a static magnetic field. The eddy current in the static magnetic field is subjected to Lorentz force, which causes vibration that is proportional to the magnitude of the internal current in the media. The system was tested to measure a current carrying wire as well as a living hamster. The resulting vibration is at the frequency of the oscillating magnetic field, which was easily detected using microphone since it is in the range of audible frequency.

In 1994, a complete theoretical model of magnetoacoustic imaging for bioelectric current was published by Roth et al. (1994). The theoretical model is based on the fundamental equation of continuum mechanics and electromagnetism where feasibility measurement of bioelectric current in vitro and in vivo is discussed. While in vitro imaging is very promising, Roth et al. (1994) concluded that in vivo magnetoacoustic signal that is generated by the body may be overwhelmed by the ambient noise such as sounds produced by muscle contraction and fluid or gas movement and these artifacts can be reduced by increasing the frequency of the oscillating magnetic field beyond the frequency of natural body sounds.

Later in 2006, Bin He et al. improve the system that is first inspired by Towe and Islam with the development of Magneto Acoustic Tomography with Magnetic Induction (MATMI) (Li et al. 2006; Xia et al. 2007). MATMI is a two-dimensional imaging system that shares the concept of the earlier system. However, the frequency of the oscillating magnetic field used in MATMI is in the range of ultrasonic frequency. This improvement enables the system to not only reducing the body artifacts but also producing image mapping with a resolution close to sonography (Li et al. 2006; Xia et al. 2007). MATMI was tested to image wire phantom as well as real biological tissue with different conductivity in vitro. The result shows that MATMI is capable to produce a high-resolution image and is sensitive to differentiate various types of tissue with different conductivity in the image.

In 1998, Wen et al. (1997, 1998), Wen and Bennet (2000), Wen (2000) developed Hall effect imaging, a 2D magnetoacoustic imaging system that employs different magnetoacoustic approaches. In HEI, non-focused ultrasound wave and magnetic field are combined to produce Lorentz force interaction in tissue to access tissue conductivity (Wen et al. 1997, 1998; Wen and Bennet 2000). Propagation of ultrasound wave inside the breast tissue will cause ionic charges in the breast tissue to move at high velocity due to the back and forth motion of the wave (Wen et al. 1997, 1998; Wen and Bennet 2000). Moving charges in the presence of magnetic field will experience Lorentz force. Lorentz force separates the positive and negative charges, producing an externally detectable voltage (Wen et al. 1997, 1998; Wen and Bennet 2000) that can be collected using a couple of skin electrodes (Wen et al. 1998). HEI was first tested to image a phantom made of polycarbonate that was immersed in saline solution. Later, it was tested to image biological tissue. A series of experimental studies on HEI showed that the resulting voltage was linearly proportional to the magnetic field strength and the ultrasound-induced velocity of the ionic particle.

A further study by Su et al. (2007) had improved HEI's setup when a focused ultrasound transducer was used to focus the sound wave at a focal point. This was to prevent high attenuation from occurred in tissue via ultrasound beam localization. Beam localization allowed the generation of Lorentz force interaction only at the focal point to maximize the interaction effects and increase the resulting voltage value. The ultrasound probe was attached to a 1-mm step size motor so that scanning can be done by moving the focused transducer and then 2D image can be generated. As a result, a better voltage value was obtained for the profile assessment of tissue.

Based on the review, it can be concluded that previous magnetoacoustic imaging manipulated magnetoacoustic interaction for the bioelectric profile assessment of tissue only. The output of ultrasound wave that was initially used to stimulate tissue particle motion was ignored, though it contained valuable information with regard to tissue mechanical properties. Hence, this study employs the concept of hybrid magnetoacoustic measurements, which considers the acoustic and bioelectric outputs to improve the existing breast cancer detection method.

6.1.5 Theory of Lorentz Force-Based Magnetoacoustic Imaging

Theoretically, magnetoacoustic imaging manipulates the interaction between the ultrasound wave and magnetic field in a current carrying media. Considering an ion in a breast tissue sample with charge q, a step change in conductivity σ and

Fig. 6.1 Formation of magnetoacoustic voltage at tissue interfaces.
a Ultrasound wave packet.
b Step change in conductivity and density in HMM measurement chamber due to the presence of breast sample.
c Conductivity and density gradient at tissue interfaces.
d Magnetoacoustic voltage at tissue interfaces

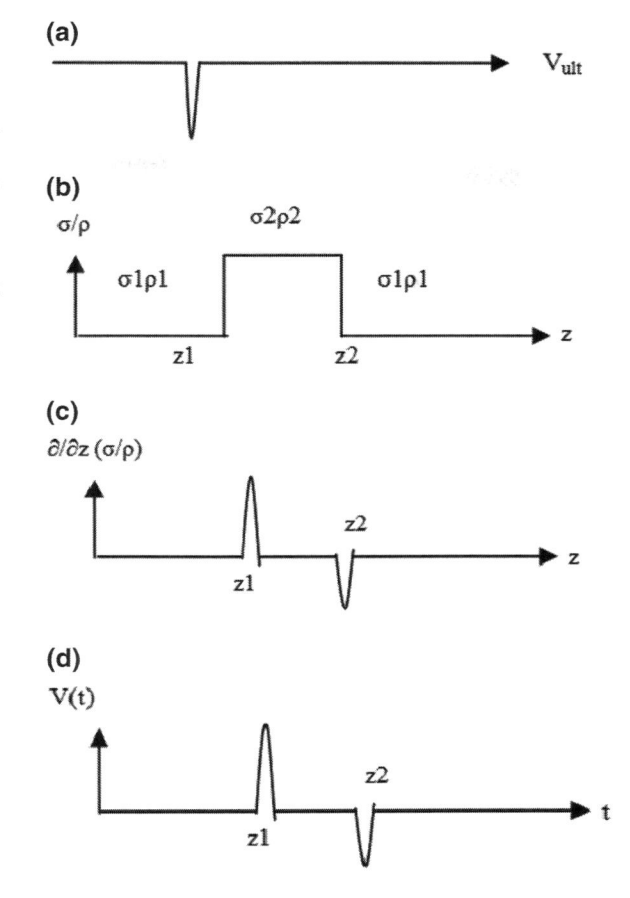

density ρ occurs between positions z_1 and z_2 in the HMM measurement chamber as shown in Fig. 6.1b due to the presence of oil and breast tissue sample.

The longitudinal motion of an ultrasound wave in z direction (Fig. 6.1a) will cause the ion to oscillate back and forth in the medium with velocity $\mathbf{V_0}$. In the presence of constant magnetic field $\mathbf{B_0}$ in y direction, the ion is subjected to the Lorentz force (Wen et al. 1997, 1998; Wen and Bennet 2000).

$$\mathbf{F} = q[\mathbf{v}_z \times \mathbf{B_0}] \tag{6.1}$$

From (6.1), the equivalent electric field is (Wen et al. 1997, 1998; Wen and Bennet 2000)

$$\mathbf{E_0} = \mathbf{v}_z \times \mathbf{B_0}. \tag{6.2}$$

The field $\mathbf{E_0}$ and current density $\mathbf{J_0}$ oscillate at the ultrasonic frequency in a direction mutually perpendicular to the ultrasound propagation path and the magnetic field $\mathbf{B_0}$ ($\mathbf{x}$ direction). The electric current density is given by (Wen et al. 1997, 1998; Wen and Bennet 2000):

$$\mathbf{J}_0 = \sigma[\mathbf{v}_z \times \mathbf{B_0}] \tag{6.3}$$

Finally, the magnetoacoustic voltage, V, across measurement electrodes a and b due to $\mathbf{J_0}$ can be calculated by the following (Su et al. 2007; Wen 2000; Zeng et al. 2010; Renzhiglova et al. 2010):

$$\mathbf{V} = \iiint [(\mathbf{v_z} \times \mathbf{B}) \cdot \mathbf{J}_{ab}/\mathbf{I}]\mathrm{d}\mathbf{V} \tag{6.4}$$

where $\mathbf{J}_{ab}$ is the current density that is induced under the electrodes surface in the breast tissue if a one ampere current, $\mathbf{I}$ is applied to the sample through the measurement electrodes (Zeng et al. 2010; Renzhiglova et al. 2010).

In addition to that, in ultrasound term, the amplitude of magnetoacoustic voltage in time domain is proportional to (Wen et al. 1998)

$$\mathrm{V}(t) \propto W \int_{\mathrm{Soundpathinitial}}^{\mathrm{Final}} \frac{\partial}{\partial z}\left[\frac{\sigma}{\rho}\right]\left[\int_0^t \partial P_z \mathrm{d}t\right]\mathrm{d}z \tag{6.5}$$

Following the equation of (6.5), gradient σ/ρ is nonzero only at interfaces z_1 and z_2 (Fig. 6.1c) giving rise to the magnetoacoustic voltage (Fig. 6.1d). The polarity of the two peaks is opposite because the σ/ρ gradients at z_1 and z_2 are in opposite direction with positive value occur at the transition of low density and conductivity area to high density and conductivity area and vice versa (Wen et al. 1997, 1998; Wen and Bennet 2000; Su et al. 2007; Wen 2000). In a uniform area within the tissue sample, the average ultrasound velocity, $\mathbf{v_0}$, is zero, and hence, no signal is observed (Wen et al. 1997, 1998; Wen and Bennet 2000).

6.2 Methodology

6.2.1 Experimental Setup

The entire experimental study was conducted in an anechoic chamber with shielding effectiveness of 18 kHz–40 GHz. Electromagnetic shielded environment is preferred to prevent external electromagnetic interference from contaminating the recorded magnetoacoustic voltage and interrupting the sensitive lock-in amplifier readings.

The HMM system consists of a 5077PR Manually Controlled Ultrasound Pulser Receiver unit, Olympus-NDT, Massachusetts, USA. The unit delivered 400 V of negative square wave pulses at the frequency of 10 MHz and PRF of 5 kHz, to 2 units of 0.125-inch standard contact, ceramic ultrasound transducers having the

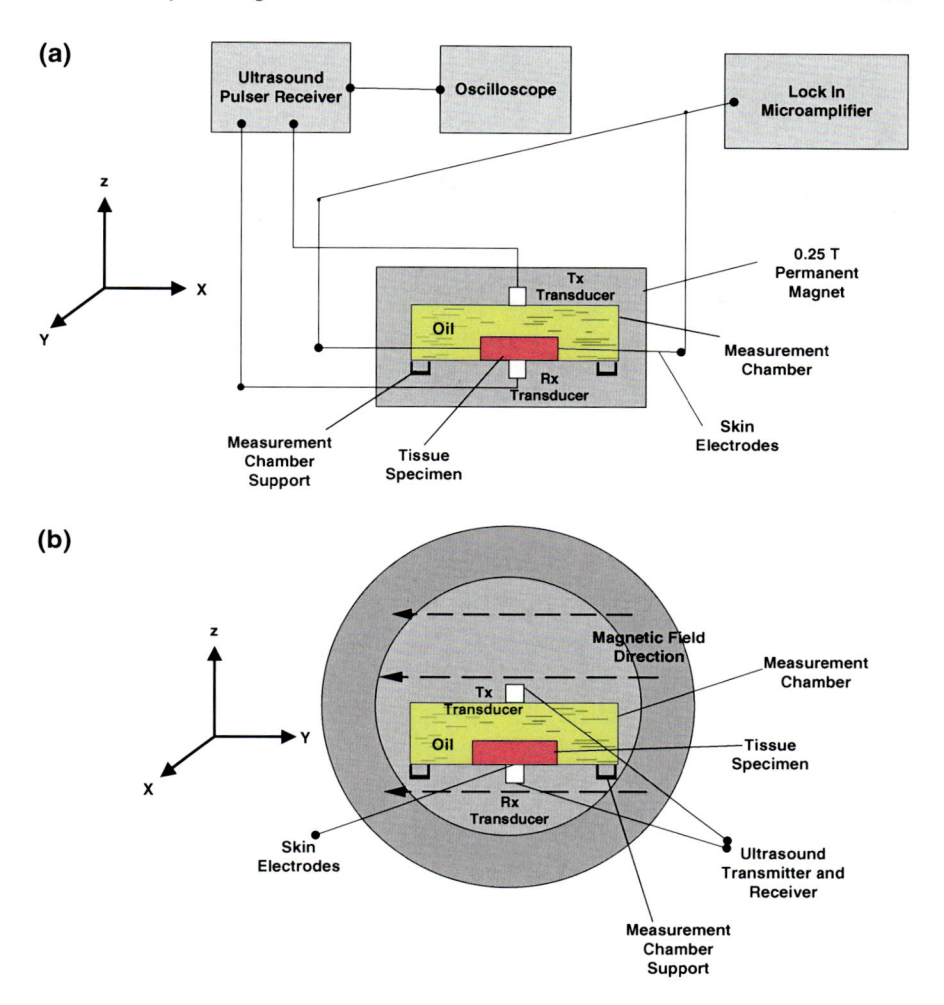

Fig. 6.2 Block diagram of the hybrid magnetoacoustic system. **a** Side view. **b** Cross-sectional view

peak frequency of 9.8 MHz. The transducers were used to transmit and receive ultrasound wave in transmission mode setting from the z direction. The pulser receiver unit was also attached to a digital oscilloscope, model TDS 3014B, Tektronix, Oregon, USA, for signal display and storage purposes.

A custom made, 15-cm height, diameter pair magnetized NdFeB permanent magnet is used to produce static magnetic field, with the intensity of 0.25T at the center of its bore. The diameter of the magnet bore is 5 cm, and the direction of magnetic field was set from the y-axis. The overall concept of HMM is shown in Fig. 6.2.

Magnetoacoustic voltage measurements were made from the x direction with respect to the measurement chamber. It was conducted using two units of custom made, ultrasensitive carbon fiber electrodes with 0.1 mm tip diameter. In general, carbon fiber electrode has been used very extensively in the in vivo (Dressman et al. 2002; Yavich and Tilhonen 2000) and in vitro including transdermal (Miller et al. 2011) bioelectrical studies of cells (Chen et al. 2011; de Asis et al. 2010) and tissue of animals (Yavich and Tilhonen 2000; Fabre et al. 1997) and humans (Crespi et al. 1995; Dressman et al. 2002; Shyu et al. 2004). On top of that, its usage in electrophysiological studies and voltammetric/amperometric analysis is perfected by its significantly less noise (Crespi et al. 1995). Furthermore, carbon fiber has a very weak paramagnetic property compared to other conventional electrodes. Due to the property, carbon fiber has been used in combination with fMRI to study the brain stimulation (Shyu et al. 2004). On top of that, carbon fiber electrode is also excellent device with greater sensitivity, selectivity and offers wide range of detectable species since its impedance can be tailored to match the sample under test (Buckshire 2008). In addition to that, many studies reported that carbon fiber electrode has the sensitivity down to 1 nV (TienWang et al. 2006; Han et al. 2004). In this study, the carbon fiber electrodes were connected to a high-frequency lock-in amplifier, model SR844, Stanford Research System, California, USA. The full-scale sensitivity of the amplifier is 100 nVrms (Stanford Research System 1997). Magnetoacoustic voltage measurement was made by touching the tip of the electrodes to the tissue in x direction.

6.2.2 Preparation of Samples

Two types of samples were used in this study. The first sample was a set of tissue mimicking gel with properties that are very close to normal breast tissue. Another sample was a set of animal breast tissues that was harvested from a group of tumor-bearing laboratory mice and its control strain shown in Fig. 6.3. The tissue mimicking gel was used in the early part of this study to understand the basic response of HMM system to linear samples before it was tested to complex samples like real tissues. The same experimental planning was also observed in previous studies (Wen et al. 1997, 1998; Wen and Bennet 2000), in which phantoms were tested to their system before they were tested on real biological tissue.

The tissue mimicking gel was prepared from a mixture of gel powder, sodium chloride (NaCl), and pure water at the right proportion to achieve the desired density and conductivity. Fifteen samples of breast tissue mimicking gel were used in this preliminary study. During the experiment, the samples were cut down to an approximately 1 cm $\times$ 1 cm size with 2-mm thickness. Thickness standardization was made using a U-shaped mold with 2-mm opening. Three random samples were chosen for a baseline density and conductivity measurement as shown in Table 6.2.

Fig. 6.3 Transgenic mice strain FVB/N-Tg MMTV-PyVT that carries high-grade invasive breast adenocarcinoma

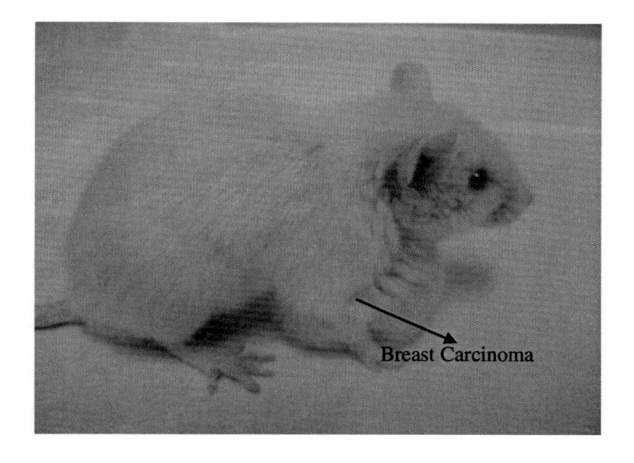

Breast Carcinoma

On the other hand, the use of animal in this study was approved by the National University of Malaysia Animal Ethics Committee. Transgenic mice strains FVB/N-Tg MMTV-PyVT 634 Mul and its control strain FVB/N were obtained from the Jackson Laboratory, USA. For the transgenic mice set, hemizygote male mice were crossed to female non-carrier to produce 50 % offspring carrying the PyVT transgene.

Transgene expression of the mice strain is characterized by the development of mammary adenocarcinoma in both male and female carriers with 100 % penetrance at 40 days of age (The Jackson Laboratory 2010; Bugge et al. 1998). All female carriers developed palpable mammary tumors as early as 5 weeks of age. Male carriers also developed these tumors at later age of onset (The Jackson Laboratory 2010; Guy et al. 1992). Adenocarcinoma that arises in virgin and breeder females as well as males is observed to be multifocal, highly fibrotic and involved the entire mammary fat pad (The Jackson Laboratory 2010; Bugge et al. 1998; Guy et al. 1992). Mice carrying the PyVT transgene also show loss of lactational ability since the first pregnancy (The Jackson Laboratory 2010). Pulmonary metastases are also observed in 94 % of tumor-bearing female mice and 80 % of tumor-bearing male mice (The Jackson Laboratory 2010; Bugge et al. 1998; Guy et al. 1992). The mice female offsprings were palpated every 3 days from 12 weeks of age to identify tumors.

Individual mouse was restrained using a plastic restrainer when the tumor diameter reached 2 cm for the transgenic mice or when it reached 18 weeks of age for normal mice. Anesthesia was performed using the Ketamin/Xylazil/Zoletil cocktail dilution. 0.2 ml of the anesthetic drug was administered intravenously from the mouse tail, and an additional of 2 ml of the drug was delivered intraperitoneally for about 2 h of sleeping time. The fur around the breast area was shaven. The mammary tissue was harvested from the mice while they were sleeping. Mice were then euthanized using drug overdose method. Excised breast specimens were cut down to an approximately 1 cm × 1 cm square shape with thickness of 2 mm immediately after the surgery to maintain the tissue

Fig. 6.4 Excised normal
breast tissue samples

Fig. 6.5 Excised cancerous
breast tissue samples

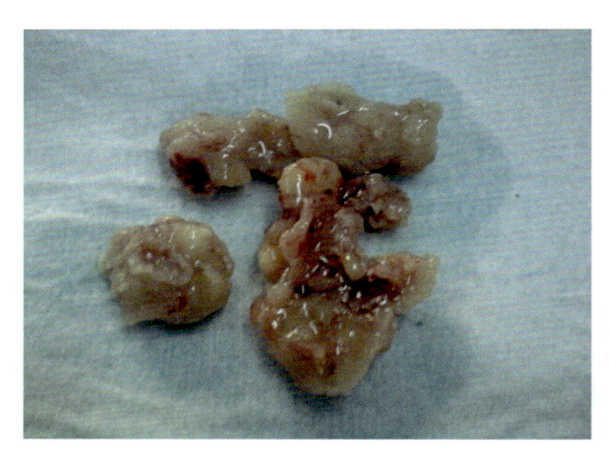

Table 6.1 Weight variations
for normal and cancerous
breast tissue samples

No.	Tissue group	Weight variation mean $\pm$ std dev (g)
1	Cancerous tissue	0.257 ± 0.03
2	Normal tissue	0.225 ± 0.02

physiological activities. The tissue was carefully trimmed down to the required thickness, and the standardization was made using a custom made U-shaped mold with 2-mm opening. A total of 24 normal and 25 cancerous breast tissue specimens were used in this study. Figures 6.4 and 6.5 show the excised breast tissue specimens. In addition to that, variation in tissue weight for normal and cancerous tissue group is presented in Table 6.1.

The overall process of trimming down after excision took an average time of 6 min, and the samples were immediately immersed in measurement chamber for scanning to maintain their physiological activities. The tissue position in

Table 6.2 Baseline conductivity and density measurement result

Samples	Conductivity (S/m)	Density (kg/m^3)
Normal mice breast	0.239	1,121
Cancerous mice breast	0.547	1,319
Gel	0.270	1,114

measurement chamber was fixed using a nylon fiber. Previous literature had shown that conductivity measurement is possible to be performed on excised tissue even though the tissue physiological states changes as a function of time following excision or death (Chaudary et al. 1984; Sha et al. 2002). This is because, the termination of blood perfusion after excision leads to the changes in ion distribution between inter- and extracellular spaces (Haemmerich et al. 2002). More specifically, cessation of sodium–potassium pump activity will lead to the depolarization of the cell that results in inflow of sodium and water, which causes cell swelling (Haemmerich et al. 2002; Lambotte 1986). Furthermore, the influx of Ca^{2+} leads to swelling and rupture of mitochondria that causes cell death.

However, the time taken for those changes to be observed differs according to few factors such as temperature and tissue types (Surowiec et al. 1985; Geddes and Baker 1967). Previous studies showed that conductivity measurement is almost constant in the first hour after the tissue sample had been excised (Chaudary et al. 1984) and measurement is still possible to be made within 4 h (Surowiec et al. 1988). In HMM, the excision of mice breast samples from its domain was done while the mice were sleeping and euthanasia was only performed at the end of the excision process. After excision, the sample was trimmed to the required size and immediately scanned. The time taken from excision to scanning took an average of 6 min and based on the literature presented, it is confirmed that the original conductivity of the breast tissue is retained during the scanning process.

Later, three random samples were chosen for a baseline density and conductivity measurement. Table 6.2 shows the mean baseline conductivity and density value for every group.

6.2.3 Experimental Data Collection

The data collection stage comprises two simultaneous measurements, which are HMM ultrasound measurement and HMM magnetoacoustic voltage measurements.

In ultrasound measurement, specimens were immersed in oil that was located between the ultrasound transmitter and receiver (Mohamad Salim et al. 2010) and its position was fixed using a nylon fiber. The ultrasound transmitter emitted 9.8-MHz ultrasound wave in transmission mode. The transmission mode approach gives some advantages including less complicated data and less noise (Landini and Sarnelli 1986). The distance between the ultrasound transmitter and receiver was set constant to 6 mm. The ultrasound analysis was started and performed at a constant temperature of 21 °C using the insertion loss method described elsewhere previously (Edmonds and Mortensen 1991; Landini and Sarnelli 1986; Mohamad Salim et al. 2010). Sonification was conducted from the z direction. Vegetable oil was used as medium for ultrasound propagation to prevent any leakage current from contaminating the measurement chamber and interfering with the HMM magnetoacoustic voltage output (Su et al. 2007).

A total of 15 gel samples were used in the early part of this study, and the measurement was taken twice for every gel sample. In addition to that, 24 normal and 25 cancerous mice breast tissue samples were also used. Measurement was repeated for five times for every biological sample at any random position on the sample surface.

Additionally, the magnetoacoustic voltage measurement was made by touching the tissue surface from the x direction using the carbon fiber electrodes. The electrodes were attached to the SR844 lock-in amplifier for signal detection and recording. The electrodes tip is the only contact point between the tissue and the detection circuit. The electrodes were manually hold to touch the tissue surface from the x direction with minimal pressure to prevent tissue dislocation from its initial position as well as to prevent measurement instability due to electrodes pressure since conductivity measurement is very sensitive to skin electrodes pressure variations. The input impedance of the lock-in amplifier was set to 1 MΩ, while the time constant was set to 3 ms. However, the experimental reading was updated every 1 s since the amplifier requires a few time constant cycle to stabilize the output reading. The recorded reading of the amplifier was equal to the average voltage of the first and second peak signals. In this study, the lock-in amplifier functions as a high precision voltage reader and a filter that detects signal as low as 100 nV at 9.8 MHz and eliminate other surrounding noises.

The gain of the SR844 lock-in amplifier is calculated by the following equation:

$$\text{Gain} = \text{Maximum scale voltage} \, (10 \, \text{V}) / \text{Sensitivity}.$$

Again, a total of 15 gel samples were used in the early part of this conductivity study, and the measurement was repeated twice for each gel. In addition to that, 24 normal and 25 cancerous breast tissue samples were used and measurement was repeated for five times for every biological tissue sample at any random position on the breast tissue surface at one measurement side (side 1). After the fifth measurement, the tissue orientation was changed to 180° from its initial orientation and measurement was repeated again for five times (side 2).

6.2.4 Experimental Data Analysis

The experimental data analysis stage comprises the analysis of HMM ultrasound output and HMM magnetoacoustic voltage output. In general, the HMM ultrasound output requires further processing in Matlab to find the attenuation scale of the ultrasound wave for every sample via spectral analysis. The objective of processing the HMM ultrasound output is to calculate the power spectral density (PSD) of the signal. It involved the determination of frequency content of a waveform via frequency decomposition. The use of PSD as an estimate of ultrasound attenuation was reported in many studies previously (Edmonds and Mortensen 1991; Mohamad Salim et al. 2010). PSD of the ultrasound signal was plotted in Matlab. The attenuation scale was calculated by subtracting the log mean-squared spectrum of ultrasound signal propagating through the oil without tissue, by the log mean-squared spectrum of ultrasound signal propagating through the oil with tissue, the following equation (Mohamad Salim et al. 2010):

$$\text{Attenuation (dB)} = \log P_0 - \log P_s \qquad (6.6)$$

where P_s is the mean-squared spectrum of the ultrasound signal propagating in the medium with tissue/gel sample, and P_0 is the mean-squared spectrum of ultrasound signal propagating through medium without sample. Later, the attenuation scale for the gel, normal tissue group, and cancerous tissue group was exported to Microsoft Excel for statistical analysis. The statistical analysis involved the determination of mean and standard deviation for every group.

On the other hand, the recorded magnetoacoustic voltage data was statistically analyzed to find its mean and standard deviation for every sample group (gel, normal tissue, and cancerous tissue).

Both, the ultrasound attenuation value and magnetoacoustic voltage value were fed to an artificial neural network (ANN) for breast cancer classification.

6.2.5 Development of Artificial Neural Network

In this study, an ANN with three inputs was developed for breast cancer classification. A total of 106 ultrasound data and 212 magnetoacoustic voltage data were collected during the experiment. The data were used as an input and target of the ANN. The ANN was trained using the steepest descent with momentum back-propagation algorithm in Matlab environment. The back-propagation algorithm is the most commonly used algorithm in medical computational application as experimented by many studies previously (Ibrahim 2005; Ibrahim et al. 2010). Measurement of ANN performance was made using the mean-squared error (MSE) (Ibrahim 2005; Ibrahim et al. 2010). Training is best when the ANN is capable to achieve the lowest MSE value.

In addition to that, each ANN configuration was tested using the testing group data to obtain the overall prediction accuracy at the end of each training and optimization process (Lammers et al. 2003; Subasi 2005). The aim is to measure the performance of the ANN architecture after each optimization. Using this method, the network was trained using a part of the data and the remainder was assigned as the testing and validation data.

The final process of development is the validation process. In this stage, a set of data that never been used during ANN training was introduced into the ANN for

Table 6.3 Details of ultrasound signals recorded by HMM

No.	Ultrasound signal	Quantity
1	Oil medium only	21
2	Oil medium with gel	30
3	Oil medium with normal breast tissue	106
4	Oil medium with cancerous breast tissue	106

classification. Validation process is very crucial as an assessment of ANN over-fitting. Overfitting is usually indicated by low-performance accuracy of ANN to validation data. This explains that the developed ANN is not able to generalize its input and requires new training and optimization process.

6.3 Experimental Result and Discussion

6.3.1 HMM Ultrasound Output

The total number of ultrasound signals that were recorded in this experiment is presented in Table 6.3.

The signals were then further processed and analyzed in Matlab to find the attenuation level of the sound wave as it propagates through the gel and the breast tissue. Attenuation is the weakening of sound wave that is characterized by the reduction in amplitude and intensity as the wave propagates through tissue (Szabo 2004; Hendee and Ritenour 2002). It encompasses the absorption as it travels and the reflection and scattering as it encounters tissue interface and heterogeneous tissues (Szabo 2004). Attenuation of ultrasound is dependent to its frequency (Edmonds and Mortensen 1991; Landini and Sarnelli 1986) as higher-frequency ultrasound experiences more attenuation as compared to lower-frequency ultrasound. In this study, the attenuation of ultrasound propagating through a material in a medium was calculated using the insertion loss method (Edmonds and Mortensen 1991; Landini and Sarnelli 1986). In the insertion loss method, the attenuation of material under test is determined by subtracting the energy of ultrasound traveling through the medium with the energy of ultrasound traveling through the medium with material under test.

Figure 6.6 shows the example of ultrasound data that was recorded in this study. The signal was converted to frequency domain using the fast Fourier transform (FFT) algorithm. After signal conditioning, the mean-squared spectrum of the ultrasound signal was determined, and finally, the PSD was calculated by converting the mean-squared spectrum to its corresponding log value. An example of a PSD plot of the ultrasound signal is shown in Fig. 6.7.

The processes described above were repeated for all ultrasound signals according to their group. Then, the corresponding mean and standard deviation of attenuation for each group were calculated. The final attenuation value was

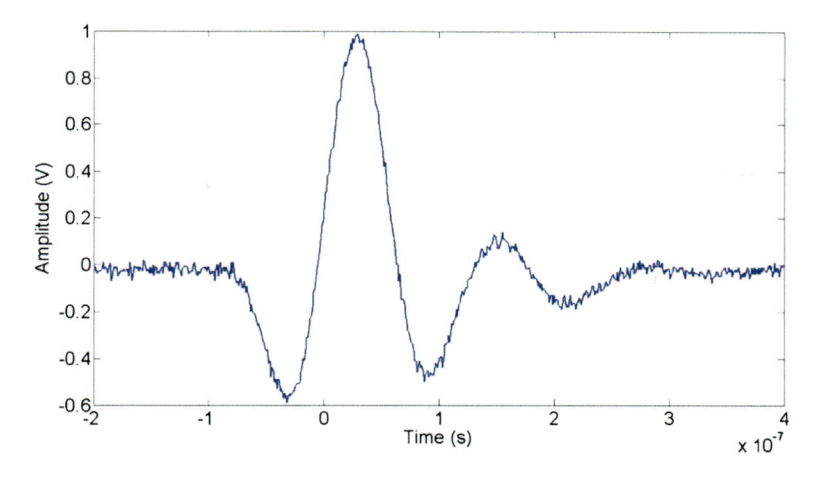

Fig. 6.6 One-dimensional ultrasound wave recorded by HMM system

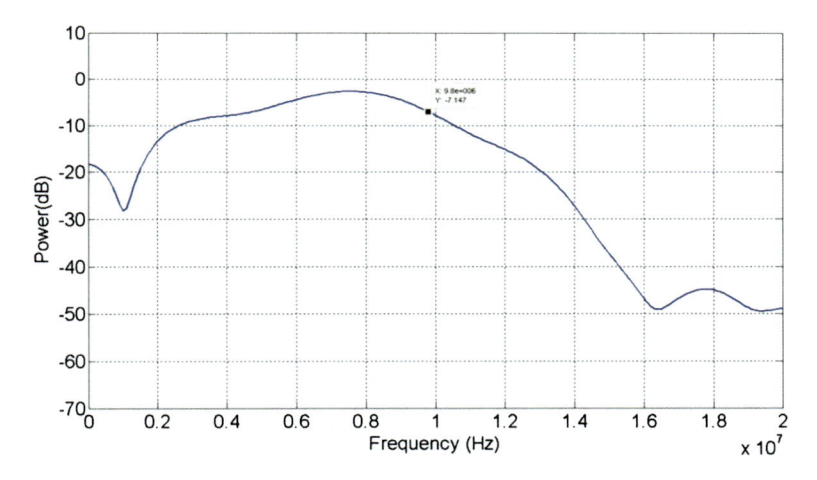

Fig. 6.7 PSD of the recorded ultrasound signal

calculated by subtracting the mean of PSD of the oil medium at 9.8 MHz by the PSD of the oil medium with the material under test (gel, normal tissue, and cancerous tissue) at 9.8 MHz.

Table 6.4 shows the final result of ultrasound attenuation scale for each group.

Absorption, scattering, and reflection are the processes that contribute to ultrasound energy attenuation in tissue. Often, the resulting attenuation is a collective result of the three processes. However, many studies reported that absorption is the most dominant factor contributing to attenuation of ultrasound wave in biological tissue via the relaxation energy loss (Kremkau 2002; Johnson et al. 2007; Berger et al. 1990).

Table 6.4 Attenuation scale of hybrid magnetoacoustic method

Type of sample	Attenuation scale (dBmm^{-1}) mean $\pm$ std dev
Tissue mimicking gel	0.501 $\pm$ 0.440
Normal breast tissue	2.329 $\pm$ 1.103
Cancerous breast tissue	1.760 $\pm$ 1.080

From the result, it can be observed that the tissue mimicking gel attenuates 0.501 ± 0.440 dB mm^{-1} of ultrasound energy, which is very small compared to the attenuation of real biological tissue group. The tissue mimicking gel is an ideal representation of tissue with linear behavior. In this study, the gel is designed to have the same density with normal tissue but to differ in its internal structure. The gel structure is homogenous, while the real breast tissue is heterogenous. The homogenous gel structure prevents energy losses due to scattering and reflection inside the gel. Hence, more ultrasound energy is likely to be preserved, and the observed standard deviation of the attenuation is also smaller compared to real biological tissue due to the structural homogeneity. However, major factor of attenuation in gel is still contributed by the absorption process.

From the real biological tissue result, it is observed that normal breast tissue attenuates ultrasound energy at the highest rate followed by the cancerous breast tissue. Similar to the gel, absorption process is a major factor that contributes to attenuation in real breast tissue samples. In general, breast tissue can be regarded as an arrangement of elastic and viscous components that consist of an aggregate of cells suspended by a viscous extracellular matrix (Kelley and McGough 2009). Extracellular matrix is usually modeled as aqueous solution of elastic polymers, which posses both solid and viscous properties (Alberts et al. 2002; Kelley and McGough 2009; Lim et al. 2000). Individual cells are modeled as elastic membranes containing viscous cytoplasm (Alberts et al. 2002; Lim et al. 2000). Overall, the resulting structure is viscoelastic materials (Alberts et al. 2002; Kelley and McG-ough 2009). In the case of breast tissue, normal breast is considered to be visco-elastic. Cancerous breast tissue that is used in this study is highly fibrotic resulting in a more elastic and denser tissue (The Jackson Laboratory 2010; Bugge et al. 1998; Guy et al. 1992; Provenzano et al. 2008). The cellular arrangements of normal and cancerous breast tissue in this study are shown in Figs. 6.8 and 6.9. The figure shows that normal cells are arranged in low density compared to the cancerous cells. Furthermore, cancerous cells have abundant protein fiber and blood vessels that further increase its density. Higher density tissue has higher inertia and resists displacement and acceleration that are caused by the ultrasound wave at its resting state (Kremkau 2002; Hendee and Ritenour 2002; Norton and Karczub 2003). Hence, it possesses very small displacement when induced by the ultrasound and quickly returns to its original equilibrium position (Hendee and Ritenour 2002; Norton and Karczub 2003). Compared to a more elastic cancerous tissue, visco-elastic molecules in normal tissue have more freedom of motion and are capable to have larger displacement and longer oscillation when induced with ultrasound. Hence, when a sound wave travels through a viscoelastic media like normal tissue,

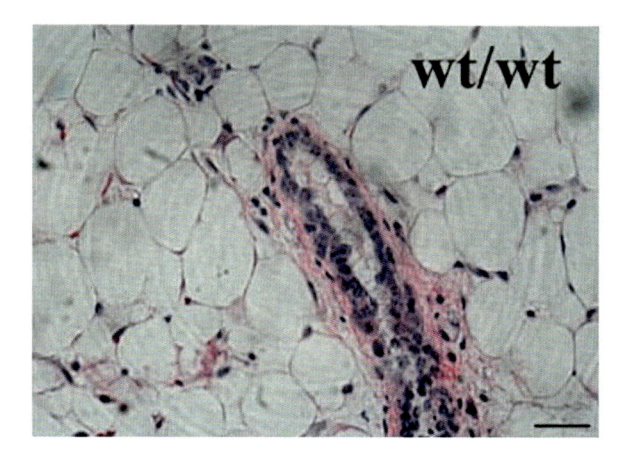

Fig. 6.8 Morphology of mammary gland from a 10-week-old normal mice model at 25 μm (Provenzano et al. 2008)

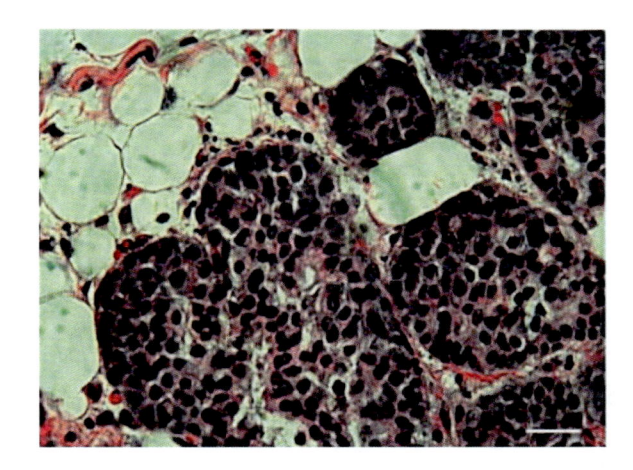

Fig. 6.9 Morphology of a high-density cancerous mammary gland of a MMTV-PyVT mice model with abundant protein fibers at 25 μm scale (Provenzano et al. 2008)

molecules vibrate for a longer period of time. This vibration requires energy that is provided to the medium by the ultrasound source. During the vibration, as the molecules attain the maximum displacement from their equilibrium, their motion stops and their kinetic energy are transformed into a potential energy associated with the position in compression zone. From this position, the molecules begin to move into the opposite direction and the potential energy is gradually transformed into kinetic energy. The conversion of kinetic to potential energy is always accompanied by energy dissipation especially when it involves larger displacement and longer oscillation such as in normal tissue. Therefore, the energy of ultrasound beam is reduced in a higher rate as it passes the viscoelastic normal tissue compared to the elastic cancerous tissue (Hendee and Ritenour 2002; Norton and Karczub 2003).

Another factor of attenuation is the reflection process that occurs at the interface between the oil and tissue. The amount of reflection at the interface is determined by the acoustic impedance difference between the oil and its adjacent medium. The average reflection coefficient that is calculated at the oil–tissue interface varied between 0.725 and 3.35 % with the highest reflection occurred at the oil–cancerous tissue interface and the lowest reflection occurred at the oil–normal tissue interface with the acoustic impedance z, as 1.45 MRayls for oil (Johnson-Selfridge and Selfridge 1985), 1.72 MRayls for normal tissue, and 2.1 MRayls for highly fibrotic cancerous tissue.

In addition to that, the heterogeneity of the breast tissue encourages further energy losses due to scattering (Kremkau 2002; Hendee and Ritenour 2002; Norton and Karczub 2003). In cancerous tissue for instance, the structural heterogeneity is contributed not only by the cellular difference, but also by the differences in the tumor region. The tumor center usually comprises necrotic tissue area that is more homogenous and becomes more heterogenous toward the outer tissue boundary. In some cases, the necrotic area is composed of an island of fluid-like structure that is characterized by its low attenuating properties but has ten times the conductivity of normal breast tissue (Wen et al. 1997, 1998; Wen and Bennet 2000). This complex cellular and tissue heterogeneity had caused the standard deviation of the calculated attenuation scale to be high for both tissue groups.

The obtained result also shows that the attenuation scale of normal and cancerous tissue is overlapping, making the detection of breast cancer using acoustic properties alone to be difficult. This result is also in agreement with a previous study that reported on the attenuation of breast cancer in human breast tissue (Edmonds and Mortensen 1991; Landini and Sarnelli 1986).

6.3.2 HMM Magnetoacoustic Voltage Output

The magnetoacoustic voltage output measurements involve detection of as low as 0.1 µV signal at the frequency of 9.8 MHz using the carbon fiber electrodes and lock-in amplifier in the x direction. The total number of magnetoacoustic voltage signals that were recorded in this experiment is presented in Table 6.5.

The magnetoacoustic voltage reading for each group was statistically analyzed to find its mean and standard deviation. The result is presented in Table 6.6.

From the table, it is observed that cancerous tissue group produces the highest magnetoacoustic voltage range, followed by the gel group and finally the normal tissue group. This is due to a few reasons that include ultrasound attenuation level in the tissue and the conductivity of the tissue. In general, it is noted from the ultrasound measurement result that normal tissue group attenuates the highest ultrasound energy. In the case of large amount of attenuation, the sound energy that is left to move the particles in the tissue decreases. Hence, the resulting particle velocity is reduced and consequently, the value of magnetoacoustic voltage is also lower. In addition to that, the conductivity of cancerous tissue group is also higher than the normal and the gel group. Therefore, high conductivity factor contributes further to increase the value of magnetoacoustic voltage of cancerous tissue. This result proves that the resulting magnetoacoustic voltage is

Table 6.5 Total number of magnetoacoustic voltage signal recorded by HMM

No.	Magnetoacoustic voltage signal	Quantity
1	Tissue mimicking gel	30
2	Normal breast tissues (1 and 2)	212
3	Cancerous breast tissues (1 and 2)	212

Table 6.6 Magnetoacoustic voltage of HMM at 9.8 MHz

Magnetoacoustic voltage	Value (μV) mean $\pm$ std dev
Tissue mimicking gel	0.56 ± 0.21
Normal breast tissues (1 and 2)	0.42 ± 0.16
Cancerous breast tissues (1 and 2)	0.80 ± 0.21

Table 6.7 Conductivity and density data of HMM specimens

Samples	Conductivity S/m	Density kg m^{-3}	σ/ρ
Normal mice breast	0.239	1,121	2.13e^{-4}
Cancerous mice breast	0.547	1,319	4.14e^{-4}
Phantom	0.27	1,114	2.42e^{-4}

not absolutely related to the tissue conductivity, but also weighted by the tissue density and uniformity that influenced the ultrasound attenuation level as explained by Eq. (6.5) (Wen et al. 1997, 1998; Wen and Bennet 2000).

$$V(t) \propto w \int_{\text{Soundpath initial}}^{\text{Final}} \frac{\partial}{\partial z}\left[\frac{\sigma}{\rho}\right]\left[\int_0^t \partial P_z dt\right] dz \qquad (6.7)$$

Table 6.7 summarizes the measured conductivity and density of HMM specimens along with the calculated σ/ρ as shown in Eq. (6.7) for comparison. The measurement results are in agreement with the experimental result in Table 6.7. The calculated σ/ρ value is the highest for the cancerous tissue, followed by the tissue mimicking phantom and finally the normal tissue.

From the presented results, it has been demonstrated that magnetoacoustic voltage gives information that is related to spatial information of tissue specifically the relative amplitude of conductivity changes across interface (Wen et al. 1997, 1998; Wen and Bennet 2000; Zeng et al. 2010; Renzhiglova et al. 2010). However, information inside the tissue cannot be obtained due to the oscillatory nature of ultrasound velocity that renders the signal to zero (Wen et al. 1997, 1998; Wen and Bennet 2000; Su et al. 2007).

The unique experimental result between specimen groups shows the potential of magnetoacoustic voltage parameter to be used in breast imaging.

Table 6.8 Classification result of the neural network

Data	Testing data		Validation data	
	Normal	Malignant	Normal	Malignant
Actual data	28	27	10	10
ANN result	25	25	8	10
% Group accuracy	89.29	92.59	80	100
% Total accuracy	90.94		90	

6.3.3 Artificial Neural Network (ANN) for Breast Cancer Classifications

All the ultrasound attenuation and magnetoacoustic voltage data that were collected and analyzed in this study were fed to an artificial neutral network for breast cancer classification and performance measurement of HMM system. The ANN was trained using the steepest descent with momentum back-propagation algorithm in Matlab environment. ANN with architecture of 3-2-1 (3 network inputs, 2 neurons in the hidden layer, 1 network output) with the learning rate of 0.3, iteration rate of 20,000, and momentum constant of 0.3 was used for classifications of breast cancer in this study.

The final classification performance of the optimum ANN for testing and validation data is shown in Table 6.8. The result indicates that the ANN is capable to achieve 90.94 and 90 % classification result for testing and validation data. This result shows the advantages of HMM output in providing additional bioelectric parameter of tissue instead of only acoustic properties for breast cancer diagnosis consideration. The system's high percentage of accuracy shows that the output of HMM is very useful in assisting diagnosis. This additional capability is hoped to improve the existing breast oncology diagnosis. However, the result of one-dimensional HMM can be further improved using 2D HMM where the classification is made to be image based rather than signal based.

6.4 Conclusion

At the end of this study, a hybrid tissue measurement method that is based on the combination of ultrasound and magnetism has been successfully developed. Compared to other conventional imaging methods such as MRI and mammography that manipulate only one property of tissue, the HMM system is capable to access the acoustic and electric properties of breast tissue for breast cancer evaluation. HMM produces two simultaneous outputs: the magnetoacoustic voltage that is related to tissue conductivity and the ultrasound attenuation scale that gives information related to tissue acoustic property. The system comprises a 400 V ultrasound pulser receiver unit, 9.8 MHz ultrasound transducer, 0.25T permanent magnet, 200 MHz lock-in microamplifier, and ultrasensitive carbon fiber skin electrodes that are capable to detect magnetoacoustic voltage down to 0.1 μV.

A series of in vitro experimental study to tissue mimicking phantom as well as to real mice breast tissue harvested from tumor-bearing mice model shows that HMM is competent to access the acoustic and electric properties of not only a simple tissue mimicking phantom, but also a complex real biological tissue. The results show that normal breast tissues experience higher attenuation $(2.329 \pm 1.103$ dB mm$^{-1})$ compared to cancerous tissue $(1.76 \pm 1.08$ dB mm$^{-1})$. In addition to that, mean magnetoacoustic voltage results for normal tissue and cancerous tissue group are 0.42 ± 0.16 and 0.8 ± 0.21 μV, respectively. The output of HMM was further automated to develop a breast cancer diagnosis system by employing the ANN. The ANN is defined by architecture 3-2-1, with the learning rate of 0.3, iteration rate of 20,000, and momentum constant of 0.3. The final ANN accuracies to testing and validation data are 90.94 and 90 %, respectively. The system's high percentage of accuracy shows that it is very useful in assisting diagnosis.

In practice, the developed one-dimensional HMM system is not yet capable to be used for in vivo imaging. Hence, further development of the system is very crucial so that this safe and non-ionizing imaging concept can be implemented in assisting breast diagnosis.

References

Alberts B, Johnson A, Lewis J, Raff M, Roberts K, Walter P (2002) Molecular biology of the cell, 4th edn. Garland Science, New York

Alizad A, Fatemi M, Wold LE, Greenleaf JF (2004) Performance of vibro-acoustography in detecting microcalcifications in excised human breast tissue: a study of 74 tissue samples. IEEE Trans Med Imag 23:307–312

Anderson ME, Soo MSC, Trahey GE (2000) Optimizing visualization for breast microcalcifications. In: 2000 IEEE ultrasonic symposium, IEEE, San Juan, Puerto Rico, pp 1315–1320, 22–25 Oct 2000

Arendt ML, Rudnick JA, Keller PJ, Kuperwasser C (2010) Stroma in breast development and disease. In: Seminar in cell and developmental biology, vol 21. Academic Press, London, pp 11–18

Bamber JC (1983) Ultrasonic propagation properties of the breast in ultrasonic examination of the breast. In: Jellins J, Kobayashi T (eds) Ultrasonic examination of the breast. Wiley, Chichester

Berger G, Laugier P, Thalabard JC, Perrin J (1990) Global breast attenuation: control group and benign breast diseases. Ultrason Imaging 12:47–57

Bissel MJ, Radisky D (2001) Putting tumors in context. Nat Rev Cancer 1:46–54

Buckshire MJ (2008) An overview of carbon fiber electrodes used in neurochemical monitoring. Master Thesis, University of Pittsburgh

Bugge TH, Lund LR, Kombrink KK, Nielsen BS, Holmback K, Drew AF, Flick MJ, Witte DP, Dano K, Degen JL (1998) Reduced metastasis of polyoma virus middle T antigen-induced mammary cancer in plasminogen- deficient mice. Oncogene 16(24):3097–3104

Chan JK, Magistris A, Loizzi V, Lin F, Rutgers J, Osann K, DiSaia PJ, Samoszuk M (2005) Mast cell density, angiogenesis, blood clotting, and prognosis in women with advanced ovarian cancer. Gynecol Oncol 99:20–25

Chang JM, Moon WK, Cho N, Park JS, Kim SJ (2011) Radiologist's performance in the detection of benign and malignant masses with 3D automated breast ultrasound (ABUS). Eur J Radiol 78:99–103

Chaudary SS, Mishra RK, Swarup A, Thomas JM (1984) Dielectric properties of normal and malignant human breast tissue at microwave and radiowave frequencies. Indian J Biochem Biophys 21:76–79

Chen CM, Chou YH, Han KC, Hung GS, Tiu CM, Chiou HJ (2003) Breast lesions on sonograms: computer-aided diagnosis with nearly setting-independent features and artificial neural networks. Radiology 226:504–514

Chen S, Hou H, Harnische F, Patil SA (2011) Electrospun and solution blown three dimensional carbon fiber nonwoven for application as electrodes in microbial fuel cells. Energy Environ Sci 4:1417–1421

Crespi F, England T, Ratti E, Trist DG (1995) Carbon fibre micro-electrodes for concomitant in vivo electrophysiological and voltammetric measurements: no reciprocal influences. Neurosci Lett 188:33–36

Cukierman E (2004) A visual quantitative analysis of fibroblastic stromagenesis in breast cancer progression. J Mammary Gland Biol Neoplasia 9:311–324

Cuzick J (2003) Epidemiology of breast cancer—selected highlights. The Breast 12:405–411

de Asis ED, Leung J, Wood S, Nguyen CV (2010) Empirical study of unipolar and bipolar configurations using high resolution single multi-walled carbon electrodes for electrophysiological probing of electrically excitable cells. Nanotechnology 21(12):125101

Degen M, Brellier F, Kain R, Ruiz C, Terracciano L, Orend G (2007) Tenascin-W is a novel marker for activated tumor stroma in low-grade human breast cancer and influences cell behavior. Cancer Res 67:9169–9179

Donegan WL, Spratt JS (2002) Cancer of the breast, 5th edn. Saunders, Philadelphia

Dressman SF, Peters JL, Michael AC (2002) Carbon fiber microelectrodes with multiple sensing elements for in vivo voltammetry. J Neurosci Methods 119:75–81

Dumane VA, Shankar PM, Piccoli CW, Reid JM, Forsberg F, Goldberg BB (2002) Computer aided classification of masses in ultrasonic mammography. Med Phys 29:1968–1973

Edmonds PD, Mortensen CL (1991) Ultrasonic tissue characterization for breast biopsy specimen. Ultrason Imaging 13(2):162–185

Fabre B, Burlet S, Cespuglio R, Bidan G (1997) Voltammetric detection of NO in the rat brain with an electronic conducting polymer and Nafion bilayer coated carbon fiber electrodes. J Electroanal Chem 426:75–83

Fern AJ (2007) Breast cancer treatment by focus microwave thermotherapy. Jones and Bartlette Learning, Cambridge

Flobbe K, Nelemans PJ, Kessels AGH, Beets GL, Von Meyenfeldt MF, Van Engelshoven JMA (2002) The roll of ultrasonography as an adjunct to mammography in the detection of breast cancer: a systematic review. Eur J Cancer 38:1044–1052

Fornage BD, Sneige N, Faroux MJ, Andry E (1990) Sonographic appearance and ultrasound guided fine needle aspiration biopsy of breast carcinomas smaller than 1 cm^3. J Ultrasound Med 9:559–568

Gasparini G (2001) Clinical significance of determination of surrogate markers of angiogenesis in breast cancer. Crit Rev Oncol Hematol 37:97–114

Geddes LA, Baker LE (1967) The specific resistance of biological materials—a compendium of data for engineer and physiologist. Med Biol Eng Comput 5:271–293

Glide C, Duric N, Littrup P (2007) Novel approach to evaluating breast density utilizing ultrasound tomography. Med Phys 34(2):744–753

Guy CT, Cardiff RD, Muller WJ (1992) Induction of mammary tumors by expression of polyomavirus middle T oncogene: a transgenic mouse model for metastatic disease. Mol Cell Biol 12(3):954–961

Haemmerich D, Ozkan OR, Tsai JZ, Staelin ST, Tungjitkusolmun S, Mahvi DM, Webster JG (2002) Changes in electrical resistivity of swine liver after occlusion and post mortem. Med Biol Eng Comput 40:29–33

Han X, Wang CT, Bai J, Chapman ER, Jackson MB (2004) Transmembrane segments of syntaxin line the fusion pore of Ca^{2+}-triggered exocytosis. Science 304:289–292

Hendee WR, Ritenour ER (2002) Medical imaging physics, 4th edn. Wiley Liss, New York

Horsch K, Giger ML, Vyborny CJ, Venta LA (2004) Performance of computer aided diagnosis in the interpretation of lesion on breast sonography. Acad Radiol 11:272–280

Ibrahim F (2005) Prognosis of dengue fever and dengue hemorrhagic fever using bioelectrical impedance. Ph.D. Thesis, Department of Biomedical Engineering, Faculty of Engineering, University of Malaya

Ibrahim F, Faisal T, Mohamad Salim MI, Taib MN (2010) Non invasive diagnosis of risk in dengue patients using bioelectrical impedance analysis and artificial neural network. Med Biol Eng Comput 48:1141–1148

Islam MR, Towe BC (1988) Bioelectric current image reconstruction from magneto-acoustic measurements. IEEE Trans Med Imaging 7(4):386–391

Johnson SA, Abbott T, Bell R, Berggren M, Borup D, Robinson D, Wiskin J, Olsen S, Hanover B (2007) Non-invasive breast tissue characterization using ultrasound speed and attenuation—in vivo validation. In: Acoustical imaging. Springer, Berlin, pp 147–154

Johnson-Self-ridge P, Selfridge RA (1985) Approximate materials properties in isotropic materials. IEEE Trans Ultrason Ferroelectr Freq Control SU-32:381

Jossinet J (1996) Variability of impedivity in normal and pathological breast tissue. Med Biol Eng Comput 34:346–350

Jossinet J, Lobel A, Michoudet C, Schmitt M (1985) Quantitative technique for bio-electrical spectroscopy. J Biomed Eng 7:289–294

Kelley JF, McGough RJ (2009) Fractal ladder models and power law wave equation. J Acoust Soc Am 126(4):2072–2081

Kremkau FW (2002) Diagnostic ultrasound principles and instruments, 6th edn. Saunders, Philadelphia

Kuo WJ, Chang RF, Moon WK (2002) Computer aided diagnosis of breast tumors with different US system. Acad Radiol 9:793–799

Lambotte L (1986) Cellular swelling and anoxic injury of the liver. Eur Surg Res 18:224–229

Lammers RI, Hudson DL, Seaman ME (2003) Prediction of traumatic wound infection with a neural network-derived decision model. Am J Emerg Med 21(1):1–7

Landini L, Sarnelli S (1986) Evaluation of the attenuation coefficient in normal and pathological breast tissue. Med Biol Eng Comput 24:243–247

Li X, Xu Y, He B (2006) A phantom study of magnetoacoustic tomography with magnetic induction (MAT-MI) for imaging electrical impedance of biological tissue. J Appl Phys 99(6):066112

Lim CT, Zhou EH, Quek ST (2000) Mechanical model for living cell: a review. J Biomech 39:195–216

Locasale JW, Cantley LC (2010) Altered metabolism in cancer. BMC Biol 88:88

Miller PR, Gittard SD, Edward TL, Lopez DM, Xiao X (2011) Integrated carbon fiber electrodes within hollow polymer microneedles for transdermal electrochemical sensing. Biomicrofluidics 5–013415:1–14

Mohamad Salim MI, Ahmmad SNZ, Rosidi B, Ariffin I, Ahmad AH, Supriyanto E (2010) Measurements of ultrasound attenuation for normal and pathological mice breast tissue using 10 MHz ultrasound wave. In: Proceeding of the 3rd WSEAS international conference on visualization, imaging and simulation (VIS'10), WSEAS, Faro, Portugal, pp 118–122, 3–5 Nov 2010

Moinfar F (2007) Essentials of diagnostic breast pathology: a practical approach. Springer, New York

Norton SJ (2003) Can ultrasound be used to stimulate nerve tissue? Biomed Eng Online 2:1–9

Norton M, Karczub D (2003) Fundamentals of noise and vibration analysis for engineer, 2nd edn. Cambridge Press, Cambridge

Orimo A, Gupta PB, Sgroi DC, Arenzana-Seisdedos F, Delaunay T, Naeem R (2005) Stromal fibroblast presents in invasive human breast carcinomas promotes tumor growth and angiogenesis through elevated SDF-1/CXCL 12 secretions. Cell 121:335–348

Provenzano PP, Inman DR, Eliceiri KW, Knittel JG, Yan L, Rueden CT, White JG, Keely PJ (2008) Collagen density promotes mammary tumor initiation and progression. BMC Med 6(11):1–15

Renzhiglova E, Ivantsiv V, Xu Y (2010) Difference frequency magneto-acousto-electrical tomography (DF-MAET): application of ultrasound induced radiation force to imaging electrical current density. IEEE Trans Ultrason Ferroelectr Freq Control 57(11):2391–2402

Riordan J (2005) Breastfeeding and human lactation, 3rd edn. Jones and Bartlett learning, Massachusetts

Roth BJ, Basser PJ, Wiksowo JP (1994) A theoretical model for magneto-acoustic imaging of bioelectric currents. IEEE Trans Biomed Eng 41(8):723–728

Sahiner B, Chan HP, Roubidoux MA, Helvie MA, Hadjiiski LM, Ramachandran A (2004) Computerized characterization of breast masses on three-dimensional ultrasound volumes. Med Phys 31:744–754

Sehgal CM, Weinstein SP, Arger PH, Conant EF (2006) A review of breast ultrasound. J Mammary Gland Biol Neoplasia 11:113–123

Sha L, Ward ER, Story B (2002) A review of dielectric properties of normal and malignant breast tissue. In: Proceeding of IEEE SoutheastCon 2002, IEEE, pp 457–462

Shankar PM, Dumane VA, Piccoli CW, Reid JM, Forsberg F, Goldberg BB (2002) Classification of breast masses in ultrasonic B mode images using a compounding technique in the Nakagami distribution domain. Ultrasound Med Biol 28:1295–1300

Shinoji M, Hancock WW, Abe K, Micko C, Casper KA, Baine RM (1998) Activation of coagulation and angiogenesis in cancer: immunohistochemical localization in situ of clotting proteins and vascular endothelial growth factor in human cancer. Am J Pathol 152(2):399–411

Shyu BC, Lin CY, Sun JJ, Sylantyev S, Chang C (2004) A method for direct thalamic stimulation in fMRI studies using a glass-coated carbon fiber electrode. J Neurosci Methods 137:123–131

Singh K, Azad T, Dev Gupta G (2008) The accuracy of ultrasound in diagnosis of palpable breast lump. JK Sci 10(4):186–188

Stanford Research System (1997) User's manual, SR844 RF lock in amplifier, USA

Stavros AT (2004) Breast ultrasound. William and Wilkins, Philadelphia

Stavros AT, Thickman D, Rapp CL, Dennis MA, Parker SH, Sisney GA (1995) Solid breast nodules: use of sonography to distinguish between benign and malignant lesions. Radiology 196:132–134

Su Y, Haider S, Hrbek A (2007) Magnetoacousto electrical tomography, a new imaging modality for electrical impedance. In: Proceeding of 13th international conference on electrical bioimpedance and the 8th conference on electrical impedance tomography IFMBE. Springer, Graz, Austria, pp 292–295, Aug 29–Sept 2007

Subasi A (2005) Automatic recognition of alertness level from EEG by using artificial neural network and wavelet coefficient. Expert Syst Appl 28:701–711

Surowiec A, Stuchly SS, Swarup A (1985) Radiofrequency dielectric properties of animal tissue as a function of time following death. Phys Med Biol 30(10):1131–1141

Surowiec AJ, Stuchly SS, Barr JB, Swarup A (1988) Dielectric properties of breast carcinoma and the surrounding tissue. IEEE Trans Biomed Eng 35:257–263

Svensson WE (1997) A review of the current status of breast ultrasound. Eur J Ultrasound 6:77–101

Szabo TL (2004) Diagnostic ultrasound imaging: inside out. Academic Press, London

The Jackson Laboratory (2010) JAX Mice Database: MMTV-PyVT strain, USA

The W, Wilson ARM (1998) The role of breast ultrasound in breast cancer screening. A consensus statement by the European group for breast cancer screening. Eur J Cancer 34(4):449–450

Thomas HV, Reeves GK, Key T (1997) Endogenous estrogens and postmenopausal breast cancer: a quantitative review. Cancer Causes Control 8:992–998

TienWang C, Bai J, Chang PY, Chapman ER, Jackson MB (2006) Synaptotagmin–Ca^{2+} triggers two sequential steps in regulated exocytosis in rat PC12 cells: fusion pore opening and fusion pore dilation. J Physiol 570(2):295–307

Towe BC, Islam MR (1988) A magneto acoustic method for the noninvasive measurement of bioelectric current. IEEE Trans Biomed Eng 35(10):892–894

Weinstein SP, Seghal CM, Conant EF, Patton JA (2002) Microcalcifications in breast tissue phantoms visualized with acoustic resonance coupled with power Doppler US: initial observations. Radiology 224:265–269

Wen H (2000) Feasibility of biomedical application of Hall Effect imaging. Ultrason Imag 22:123–136

Wen H, Bennet E (2000) The feasibility of Hall Effect imaging in humans. In: 2000 IEEE ultrasonic symposium, San Juan, Puerto Rico: IEEE. pp 1619–1622, 22–25 Oct 2000

Wen H, Bennett E, Shah J, Balaban RS (1997) An imaging method using ultrasound and magnetic field. In: Proceeding of the 1997 IEEE ultrasonic symposium. Toronto, Ontario, IEEE, pp 1407–1410, 5–8 Oct 1997

Wen H, Shah J, Balaban RS (1998) Hall Effect imaging. IEEE Trans Biomed Eng 45:119–124

Xia R, Li X, He B (2007) Magnetoacoustic tomography of biological tissue with magnetic induction. In: Joint meeting of the 6th international symposium on noninvasive functional source of the brain and heart and the international conference on functional biomedical imaging 2007, IEEE, Hangzhou, China, pp 287–287, 12–14 Oct 2007

Yavich L, Tilhonen J (2000) In vivo voltammetry with removable carbon fiber electrodes in freely moving mice: dopamine release during intracranial self stimulation. J Neurosci Method 104:55–63

Zeng X, Liu G, Xia H, Xu X (2010) An acoustic characteristics study of magneto-acousto-electrical tomography: a new method to reconstruct current density distribution at every point of a sample. In: 2010 3rd international conference on biomedical engineering and informatics, IEEE, Yantai, pp 95–98, 16–18 Oct 2010

Zou Y, Guo Z (2003) A review of electrical impedance techniques for breast cancer detection. Med Eng Phys 25:79–90

Chapter 7
Sequential Process of Emotional Information from Facial Expressions: Simple Event-Related Potential (ERP) for the Study of Brain Activities

Nugraha Priya Utama, Khin Wee Lai, Maheza Irna Mohamad Salim, Yan Chai Hum and Yin Mon Myint

Abstract Human is socially living creature that needs to communicate with others. In direct communication, there are two ways in conveying the information: through speaking words or verbally and through facial expression, body gesture or non-verbally. The non-verbal communication is taken almost 70 % of humans' communication. Therefore, to understand how this non-verbal information is processed by the brain is quite important. In this chapter, we would like to elucidate the process of the brain in understanding the facial expression by analyzing the brain signals that correspond to emotional content of facial expression. As known, the emotion can be differentiated into the type and the level of emotion. For example, though we know that smiley face and joyful face belong to the same type of happiness, we know that the level of happiness is higher in the joyful face. Therefore, how does the brain process this kind of type and level of emotional information is the basic question that we would like to answer in this chapter. In this chapter, we explain the way we collect the data, the step-by-step process of reducing the noise in the brain signals, the way of inter-correlating the behavioral

N. P. Utama (✉) · Y. M. Myint
Faculty of Biosciences and Medical Engineering, Universiti Teknologi Malaysia, Level 3, V01, Block A, Satellite Building, Skudai-Johor, Malaysia
e-mail: utama@biomedical.utm.my

Y. M. Myint
e-mail: yinmontt@biomedical.utm.my

K. W. Lai
Biomedical Engineering Department, University Malaya, Kuala Lumpur, Malaysia
e-mail: lai.khinwee@um.edu.my

M. I. Mohamad Salim
Universiti Teknologi Malaysia, Level 3, V01, Block A, Satellite Building, Skudai-Johor, Malaysia
e-mail: maheza@biomedical.utm.my

Y. C. Hum
MIMOS Berhad, Technology Park Malaysia, 57000 Kuala Lumpur, Malaysia
e-mail: yc.hum@mimos.my

K. W. Lai et al., *Advances in Medical Diagnostic Technology*,
Lecture Notes in Bioengineering, DOI: 10.1007/978-981-4585-72-9_7,
© Springer Science+Business Media Singapore 2014

data and brain signals, how we used those data to find the location of activated brain area for processing the specific content of emotion, and finally, how we exactly find that the process of understanding the emotional information from facial expression is a sequential process; understanding the type of emotion, followed by the level of that specific emotion. This emotional process is different from that of the process of understanding the physical content of the face, like identity and gender.

7.1 Introduction

Humans are socially living creatures. They need to interact with others, and for living comfortably, communications among humans have a significant and important role in it. Generally, communications can be divided into two big separable groups, first is the verbal communications, process for conveying and for expressing meaning with the verbal language. The other one is the non-verbal communications, just like its name, it does not contain any orally meaning words, the communications without words.

If we see throughout the life, in every single thing of the daily interactions, never human can be free from this second type of communications, the non-verbal communications. Even while speaking, humans combine their words with the arm movement, facial expression, and so on, for helping them to make the information easier to be understood by the opponent. The first major scientific study of non-verbal communications, especially the facial communication, was published by Darwin in 1872. Darwin concluded that many expressions and their meanings (e.g., for astonishment, shame, fear, horror, pride, hatred, wrath, love, joy, guilt, anxiety, shyness, and modesty) are universal: '*I have endeavoured to show in considerable detail that all the chief expressions exhibited by man are the same throughout the world*' (Darwin 1872). Many researchers like Silvan Tomkins, Carrol Izard, and Paul Ekman were supporting the universality of facial expressions. But, there were also many other researchers were not supporting this universality idea. Margaret Mead, Gregory Bateson, Edward Hall, Ray Birdwhistell, and Charles Osgood were the researchers who said that the expressions and gestures are culture's variable, and they were learnt through the social interaction. They said that in many cultures, people keep smiling even when they were in lost. And Birdwhistell concluded that everything that is socially important, like the expression of emotions, must be the product of learning, so they will be different among cultures (Ekman 2003).

For us, these differences, whether the facial expressions are product of cultures or universal, are not so important. Those groups saw the emotional expression from different point of views, and of course, the conclusion would be naturally different. The group, which said that the facial expressions are the product of cultures, was seeing through the facial expression for social interaction, meanwhile the universality group was seeing it through the individual perception in single expression without bias of social faces. What we concern is the way our brain processes the emotional information, especially the facial expression; therefore, we take the universality point of view for the facial expression, an individual perception without bias of social faces. Happy expression will be seen as happy expression and angry as angry without any social disturbance.

As known, we humans can differentiate the facial expression of our opponent generally into its type and the emotional level of that expression. For example, we can easily distinguish the difference of happiness from anger, and we also can evaluate the emotional level of that expression, whether they are in totally joyful or manic condition or just the ordinary happiness. Many researchers have shown that the brain responded to emotionally charged stimuli more than that of the neutrally rated stimuli (Fredikson et al. 1995; Adolphs and Tranel 2003; Balconi and Pozzoli 2003), but less or maybe none of them explored the effect of emotional strength on the brain signals. In this chapter, we would like to answer the fundamental questions of *'How do the brains distinguish the type of emotions?'* and *'How does the brain understand the strength of the emotion from the facial expressions?'* When and which area of the brain processes these kinds of emotional information.

7.2 Selected Researches at Emotional Effect in the Brain

When we saw an agony or sadness in the face of someone, we, instinctively, will try to help, to support, to do something to relieve the sadness or the agony. This agony or sadness in faces is a kind of sign or call for help from others. Not like sadness or agony, happiness, surprise, and others facial expressions also have their own special meaning to be responded. In general, humans will react differently, depends on the facial expressions of the opponent.

Recognizing the facial expressions is one of the very skilled ability of human, even babies precociously respond to different facial expressions (Field et al. 1982). From the clinical study, even people with mobius syndrome, a disorder producing facial paralysis, are able to recognize facial expressions (Philipps et al. 1998; Calder et al. 2000). Experimental studies on normal subjects showed that when the subjects were asked to make quick judgments of emotional expressions, their reaction times in judging the emotional contents in the presented expression were equal for both familiar and unfamiliar faces. By these results, we can conclude that the process of recognizing the facial expression is a special process that independents from a structural recognition and the identity of faces.

Cells within temporal visual cortices have long been known to show robust responses to faces, which are modulated by two factors: attention and emotion. Yet, a large number of psychological studies—over past decades in the process of recognizing emotion from facial expressions—sheer diversity in their findings. Wealth neurobiological findings from experiment involving lesions, event-related potential (ERP), magnetoencephalography (MEG), positron emission tomography (PET), and functional magnetic resonance imaging (fMRI) preclude any simple summary and argue against the isolation of only few structures. Instead, it is becoming clear that recognizing facial emotion draws on multiple strategies sub-served by a large array of different brain structures. The neuropsychology of emotion has stressed the left–right brain dimension as fundamental for emotional valences (Heller et al. 1998) with right-hemispheric superiority when processing negative connotations of incoming information and left superiority for positive connotations (Schwartz et al. 1975; Reuter-Lorenz and Davidson 1981; Natale et al. 1983). Clinical study showed agreeing lateralization: depression is associated predominantly with left-hemispheric lesions, inappropriate cheerfulness with right-hemispheric lesions (Robinson 1995). Not so stressing in the left hemisphere for emotional judgments, other researchers concluded that the structures in the right hemisphere appear to be important for the normal processing of emotional and social information (Benowitz et al. 1983; Borod et al. 1985; Bowers et al. 1985). From other lesions studies, damage to the right-hemisphere somatosensory cortices (RSS) caused the impairment for recognizing six basic emotions from facial expressions (Heberlein et al. 2003).

Recent ERP studies have supported the hypothesis that the process of facial expression recognition starts very early in the brain. The 120-ms post-stimulus onset was assumed to be the first perceptive stage, in which the subject completes the 'structural codes' of face, which is thought to be processed separately from complex facial information such as emotional meaning (Lane et al. 1998; Pizzagalli et al. 1999a; Junghofer et al. 2001; Utama et al. 2009), and in addition to a 'structural code,' the existence around 170-ms post-stimulus onset was supposed of an 'expression code' implicated in the decoding of emotional facial expressions (Bruce and Young 1998; Ellis and Young 1998). Faces with emotions elicited a larger negative peak at approximately 270 ms than neutral faces over the posterior visual area (Sato et al. 2001), and the differences in peak amplitude of the brain potential were affected by the experienced emotional intensity, related to arousal and unpleasant value of the stimulus (Balconi and Pozzoli 2003; Utama et al. 2009).

7.3 Psychophysics Experiment

Action and reaction are two natural things happen for humans, and so for the brain. Response in human can be concluded as the response of the brain, especially when the response corresponds to some actions which are related to higher level of mechanisms, such as cognitive-related tasks, emotional-related tasks, and other more. Psychophysics experiment is the name for an experiment which explores the action and reaction of humans or brain under manipulated situation and condition to scientifically study their relation. This concept can be applied into many experimental applications, and so for elucidating the process of emotional information in the brain.

ERP is a study that investigates changes in brain signals correspond to the presented stimuli. In this writing, we would like to combine the ERP with

(a)

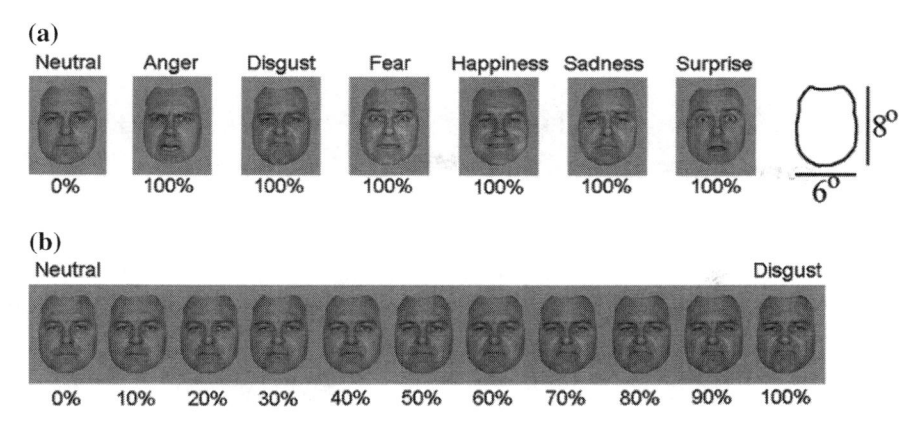

(b)

Fig. 7.1 Sample stimuli. **a** Sample of original images derived from the Ekman and Friesen collections of neutral face and the facial emotions of anger, disgust, fear, happiness, sadness, and surprise. The images were cropped with the same outline (6° × 8°). Numbers were designated 0 % as the neutral face and 100 % as the intensity of the original image of the facial emotion. **b** Transitional images from the neutral face to the emotion disgusted are indicated as the 10 % increment values

psychophysics experiment using several facial images that have been subjectively rated by normal healthy humans, based on two categories, the type and the emotional level. By the type of emotions, it means that the image was rated if it belonged to the type of facial emotions, such as happiness, sadness, disgust, fear, surprise, anger, and neutral of facial expressions, and by the emotional level, it means the image was rated into ten different levels of emotions, from one (1) to ten (10) in describing the lowest to the highest emotional level of facial expression, respectively. Special for the images of neutral facial expression, the emotional level is set to be zero (0) as default.

For this subjective rating, the original images of facial expressions were taken from Ekman and Friesen (1976). The images were black-and-white pictures of three (3) male actors, presenting happy, disgust, fear, surprise, angry, sad, and neutral faces. Each of the images was morphed from neutral face to each of those emotional types into nine steps of morphing, in 10 % of increment. Total stimuli for this experiment were 198 photos, contained 18 neutral faces and 180 emotional faces from three (3) different actors. Sample of morphing images which were used as stimulus can be seen in Fig. 7.1.

Besides the recording of subjective rating from the subjects of experiment, the brain signals of the subjects were also recorded during the whole experiments. These two different data, subjective rating and brain signals, will be inter-correlated in order to elucidate the temporal effect of facial expressions in humans' brain. Besides that, the task of rating *the presented facial expression subjectively* is one way to keep the subject stay awake and alert during the experiment.

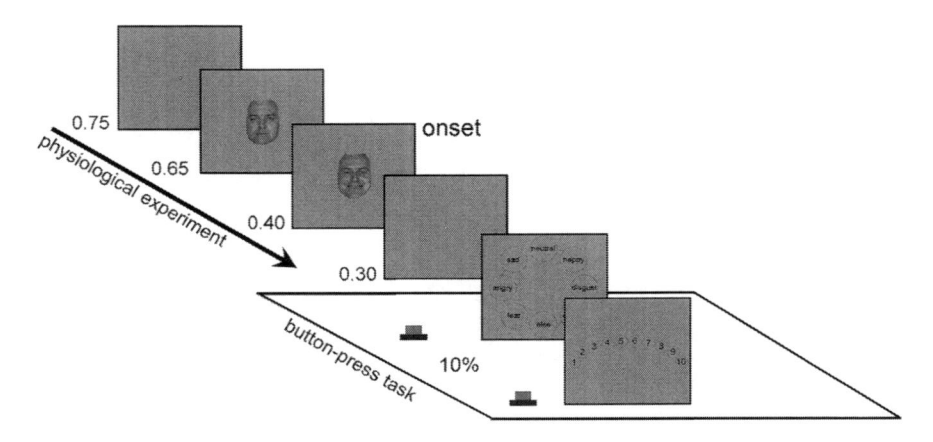

Fig. 7.2 Experimental design of psychological experiment. A trial was started by the presentation *blank gray screen* with *white dot* in the center of the screen as fixation point. A neutral face was then presented followed by the presentation of a randomly selected transitional image of one of the emotional facial expressions from the same actor. The subject was required to identify the type of emotion and to assess the intensity of each image. Presentation time [s] is shown at the *left sides* of each panel

We presented facial expressions as still photo images at the center of a CRT 21″ (1,280 × 1,024, 100 Hz) where monitor placed approximately 70 cm from the subjects, with a visual horizontal angle 4° and a vertical angle of 6°. To fixate the subjects' gaze, a white dot was presented at the center of the display during the transition between image presentations or between trials. The subject was told to stare at the fixation point, to carefully watch the stimulus and to do the press-button tasks as quickly as possible after the instruction was shown. Detail of the experimental design is shown in Fig. 7.2.

7.4 Brain Signal Recordings

Despites many medical imaging techniques can be used for recording the brain activities, electroencephalography is still the least dangerous, the less expensive, and yet the most temporally accurate for elucidating the brain activities based on its signals. Electroencephalography is a science of recording and analyzing the electrical activity of the brain. The spontaneous brain's electrical activities are recorded in the time domain from several electrodes placed on the scalp. The electrodes are then linked to an electroencephalograph, which is an amplifier connected to a mechanism that converts electrical impulses into digital data and displayed on a computer screen. The digital data or the print out of spontaneous brain's activity is called an electroencephalogram (EEG) (see Fig. 7.3).

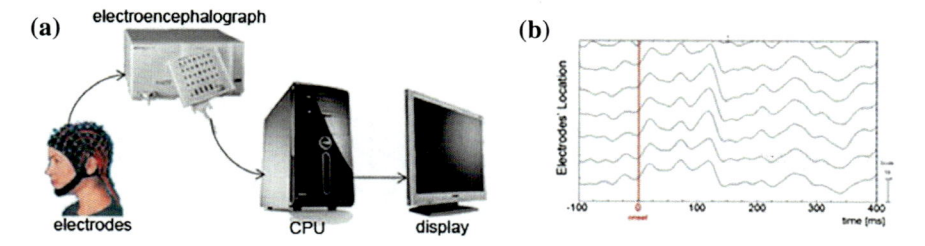

Fig. 7.3 a EEG system. **b** Samples of raw EEG

7.4.1 Brief History of Electroencephalography

Experiment by applying electricity to dead frogs' nerve trunks which induced the movement of their legs was the first demonstration to prove that information in the nervous system may be electrically transmitted. This experiment was done by Italian physiologist Luigi Galvani during the eighteenth century. In 1870, two Prussian (current: German) physicians, Gustav Theodor Fritsch and Eduard Hitzig confirmed Galvani's work by electrically stimulating areas of motor cortex of the dog's brain which caused involuntary muscular contractions of specific parts of its body. It was not until 1875, however, that the Liverpool physician, Dr. Richard Caton, became the first person to record electrical activity in the brain by placing electrodes directly on brains of vivisected rabbits and monkeys. Using a primitive measuring device known as a mirror galvanometer, in which a moving mirror was used to amplify very small voltages, he reported finding feeble currents in the cerebral cortex, the outermost layer of the brain (Finger 1994).

Electrophysiological recordings became much more fashionable after Hans Berger (1873–1941) published his human EEG in 1929. The first human EEG was recorded using electrodes (made of lead, zinc, platinum, etc.) attached to the intact skull and connected to an oscillograph. Berger made 73 EEG recordings from his fifteen-year-old son, Klaus. The first frequency he encountered was the 10-Hz range (8–12 Hz), which at first was called the Berger rhythm, currently called alpha rhythm brain wave. He reported that the brain generates electrical impulses or 'brain waves.' The brain waves changed dramatically if the subject simply shifts from sitting quietly with eyes closed (short or alpha waves) to sitting quietly with eyes opened (long or beta waves). Furthermore, brain waves also changed when the subject sat quietly with eyes closed, 'focusing' on solving a math problem (beta waves). That is, the electrical brain wave pattern shifts with attention (O'Leary 1970). The publication of Hans Berger's in 1929 changed neurophysiology forever, and because of it, he earned the recognition of 'Father of Electroencephalography.'

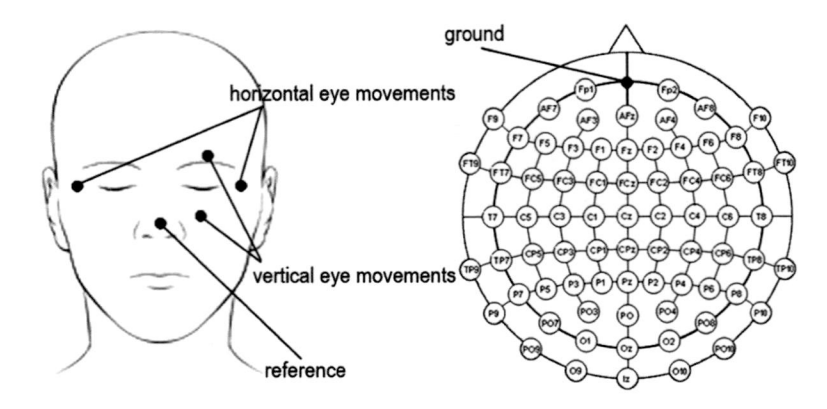

Fig. 7.4 Electrodes nomenclature and placement

7.4.2 The Electroencephalograph

The system for recording the spontaneous brain's activity is electroencephalograph. It is safe, and very few risks are associated with it. Some locations on the scalp will be cleaned up by removing the dead cells and oils before attaching the electrodes. The placement of electrodes depends on the purpose of the study. In this writing, we would like to discuss based on the American 10–10 system of electrode nomenclature and placement for electrodes positioning. It contains 73 recording electrodes plus one electrode for ground and another for system reference at nose tip. All of these electrodes' positioning is embedded in a cap. Besides those electrodes, we attached two different pairs of electrodes to record horizontal and vertical movement of the eyes. We placed these electrodes on the outer canthi of the two eyes for detecting the horizontal eye movements and on the infra- and supra-orbital ridges of the left eye for the vertical eye movements (see Fig. 7.4).

For this writing, we used amplifiers of SynAmps system (Neuroscan) with amplification of 25,000 times of the voltage between the active electrode and the reference. The amplified signal is digitized via an analog-to-digital converter, after being passed through an anti-aliasing filter. Electroencephalograms (EEGs) and electrooculograms (EOGs) for the eyes were recorded continuously with a band pass filter of 0.1–100 Hz and digitization rate of 1,000 Hz.

7.4.3 The EEG Artifacts

Although EEG is designed to record cerebral activity, it also records electrical activities arising from sites other than the brain. The recorded activity that is not of cerebral origin is termed artifact and can be divided into physiological and non-physiological artifacts. Physiological artifacts are generated from the subject him/herself and include cardiac, glossokinetic, muscle, eye movement, respiratory, and pulse artifact among many others. The EEG recording can be contaminated by numerous non-physiological artifacts generated from the immediate patient surroundings. Common non-physiological artifacts include those generated by monitoring devices, infusion pumps, electric power system, and electrode pops; spikes originating from a momentary change in the impedance of a given electrode may also contaminate the EEG record.

Severe contamination of EEG activity by the artifacts is a serious problem for EEG interpretation and analysis. The easiest way to remove the artifacts is simply rejecting contaminated EEG epochs, but it causes a considerable loss of collected information. In this study, we apply independent component analysis (ICA) to multi-channel EEG recordings and remove a wide variety of artifacts from EEG records by eliminating the contributions of artifactual sources onto the scalp sensors (Jung et al. 2000a, b). ICA-based artifact correction can separate and remove a wide variety of artifacts from EEG data by linear decomposition. The ICA method is based on the assumption that firstly, the time series recorded on the scalp are spatially stable mixtures of the activities of temporally independent cerebral and artifactual sources, secondly the summation of potentials arising from different parts of the brain, scalp, and body is linear at the electrodes, and thirdly, the propagation delays from the sources to the electrodes are negligible. The second and third assumptions are quite reasonable for EEG data. Given enough input data, the first assumption is reasonable as well. The method uses spatial filters derived by the ICA algorithm and does not require a reference channel for each artifact source. Once the independent time courses of different brain and artifact sources are extracted from the data, artifact-corrected EEG signals can be derived by eliminating the contributions of the artifactual sources.

7.4.4 Reducing the Non-physiological Artifacts

Reducing the artifacts in EEG record is an important process to be done. Before the experiment, it is highly recommended to clean up the experimental room from any electrical devices that are not related with the experiment, and it is better to do the experiment in an electronically shielded room. For the artifact from electric power supply or display monitor, applying specific digital notch filter before or after EEG recording might be the easiest and the best way to reduce its effect. For the electrode-pop and electrostatic artifact, treatment after EEG recording is the only way to reduce its effect in EEG record in this study. These kinds of artifacts are originating in electrodes; the electrode-pop artifact is caused by a drying electrode or slight mechanical instability that changes the area of electrode surface in contact with the skin, and the electrostatic artifact is caused by the movement of electrode wires between the electrode on the head and the electrode board or other objects moving in relation to the input electrode leads. We cannot prevent these kinds of artifacts to be happens, but we might be able to reduce its effect in our

EEG by always using clean electrodes and ask subjects to sit still and less moving during the recording. Besides that, it is a necessity to have stable electrodes. In this study, ICA decomposition was applied to detect the ICA components with electrode-pop and electrostatic artifacts. We rejected those components to reduce the effect of these artifacts from the EEG record. The process of eliminating these ICA components was carried out under EEGLAB toolbox in Matlab (Delorme and Makeig 2004).

7.4.5 Reducing the Physiological Artifacts

Myogenic or muscle potentials are the most common artifacts in EEG recordings. Frontalis and temporalis muscles (e.g., clenching of jaw muscles) are common causes. Generally, the potentials generated in the muscles are of shorter duration than those generated in the brain and are identified easily on the basis of duration, morphology, and rate of firing (e.g., frequency). Other common physiological artifact is eye movement. Eye movement can simulate a plausible EEG slow wave having eyeball origin. Eyeball artifacts are caused by the potential difference between the cornea and retina, which is quite large compared with cerebral potentials. When the eye is completely still, this is not a problem. But, there are nearly always small or large reflexive eye movements, which generates a potential that is picked up in the frontopolar and frontal leads. Involuntary eye movements, known as saccades, are caused by ocular muscles, which also generate electromyographic potentials. Purposeful or reflexive eye blinking also generates electromyographic potentials, but more importantly, there is reflexive movement of the eyeball during blinking which gives a characteristic artifactual appearance of the EEG.

Those two artifacts above are the focus of physiological artifact in this study. Preventing action like asking subjects to control their blinking and taking the break between the recordings might reduce the occurrence of these artifacts. Applying ICA-based artifact correction can separate and remove these artifacts from EEG data (Jung et al. 2000a, b). Several useful heuristics can be used to discriminate them. For the eye movements, they should project mainly to frontal sites with a low-pass time course. For the eye blinks, they also project to frontal sites, but they have large punctate activations. For muscle artifacts, they usually project to temporal sites with a spectral peak above 20 Hz. Sample of specific topography for artifacts can be seen at Fig. 7.5.

7.4.6 The Windowing

Brain electrical activity recordings consist of time-varying measurements of the scalp electric potential field, performed for spontaneous activity (EEG) or for ERPs. In studies of ERPs, these recordings are interpreted as being formed by a

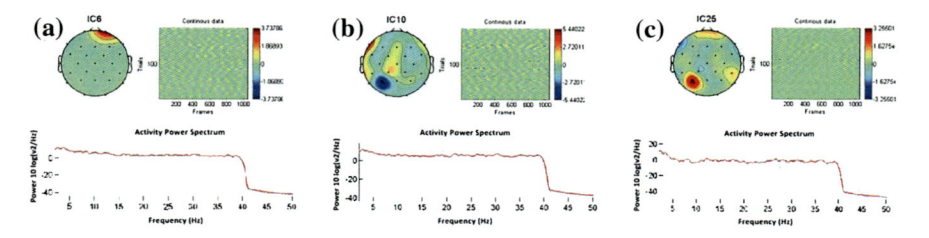

Fig. 7.5 Topography of EEG artifacts. **a** Eye artifact. **b** Muscle artifact. **c** Electrode-pop artifact

sequence of components. Each component appears as a peak or trough in the voltage versus time plot, characterized by a certain amplitude and latency value. The different components are assumed to reflect different functional states of the brain, corresponding to different stages of information processing. Therefore, the determination of the functional states and their time sequencing constitutes an important problem of electrophysiology.

7.4.7 K-Means Clustering

In studies of ERPs, the brain signals are interpreted as being performed by a sequence of components. Instead of viewing these sequences of components in the waveforms, k-means clustering technique tries to segment the brain activities into a sequence of momentary potential distribution maps (microstates). In simple way, k-means clustering technique tries to view the multi-channel records of EEG data as a sequence of microstates. Microstate is a stable topographical scalp field persists during an extended epoch or time segment (Lehmann and Skrandies 1984). Each microstate presumably reflects the different step or mode or content of information processing (Michel et al. 1992). But, we have to be careful, because the successive occurrence of microstates does not imply that brain information processing is strictly sequential. The underlying mechanism by which the brain enters a microstate with a given neuronal generator distribution may be composed of any number of sequential or parallel physiological sub-processes.

The scalp electromagnetic field reflects the source distribution in the brain. Due to the non-uniqueness of the electromagnetic inverse problem, it may occur that different source distributions produce exactly the same microstate. However, changes in the microstate are undoubtedly due to changes in the source distribution. Therefore, brain electrical activity can then be seen as a sequence of non-overlapping microstates with variable duration and variable intensity dynamics (Pascual-Marqui et al. 1995). In this technique, the EEG data are assumed to be reference free, and the entire data set at all-time points are examined simultaneously. Therefore, this technique should be applied to averaged EEG (or ERP) data after re-referenced to an average reference. Mathematical and statistical detail

Table 7.1 Sequential process of windowing with k-means clustering technique

1. Raw EEG data
2. Artifact reduction process
3. Re-referencing the 'clean' EEG data to average reference
4. Averaging the 'clean' EEG data over subjects and conditions
5. Applying k-means clustering technique
    5.1. Basic k-means clustering
        5.1.1. Checking data
        5.1.2. Setting criteria: initializing microstates
        5.1.3. Labeling each measurement to the closest microstate
        5.1.4. Computing new microstates based on labeling results
        5.1.5. Computing noise variance for the new microstates
        5.1.6. Checking for the convergence of noise variance
        5.1.7. Estimating intensity of microstates
        5.1.8. Computing the regression (R2)
        5.1.9. Cross-validated
        5.1.10. Repeat until criteria are reached

of this method can be seen at the study of Pascual-Marqui et al. (1995). Sequential process of this segmentation or windowing with k-means clustering technique can be seen at Table 7.1. Result of windowing with k-means clustering technique can be seen at Fig. 7.6.

7.4.8 Source Localization

Source localization is one issue to be answered in EEG study, and it is well known that EEG measurement does not contain enough information for the unique estimation of the electric neuronal generators. Therefore, many possible solutions exist for estimating neuronal generators, and standardized low-resolution brain electromagnetic tomography (sLORETA) is one of them (Pascual-Marqui 2002). The best thing about sLORETA is that it localizes the sources exactly for ideal conditions, but its spatial resolution decreases with depth. Because of its 'zero error' for estimating the sources, noisy measurements will produce noisy images with sLORETA estimation. In this study, on the basis of the scalp-recorded electrical potential distribution, sLORETA was used to compute the cortical three-dimensional distribution of scalp current density. Computations were made in a realistic head model (Fuchs et al. 2002), using the MNI152 template (Mazziotta et al. 2001), with the three-dimensional solution space restricted to the cortical gray matter. Anatomical labels including Brodmann areas are reported using an appropriate correction from MNI to Talairach space (Talairach and Tournoux 1988; Lancaster et al. 2000). Software for the sLORETA estimation package can be downloaded for academic purposes only from Web site of the KEY Institute for

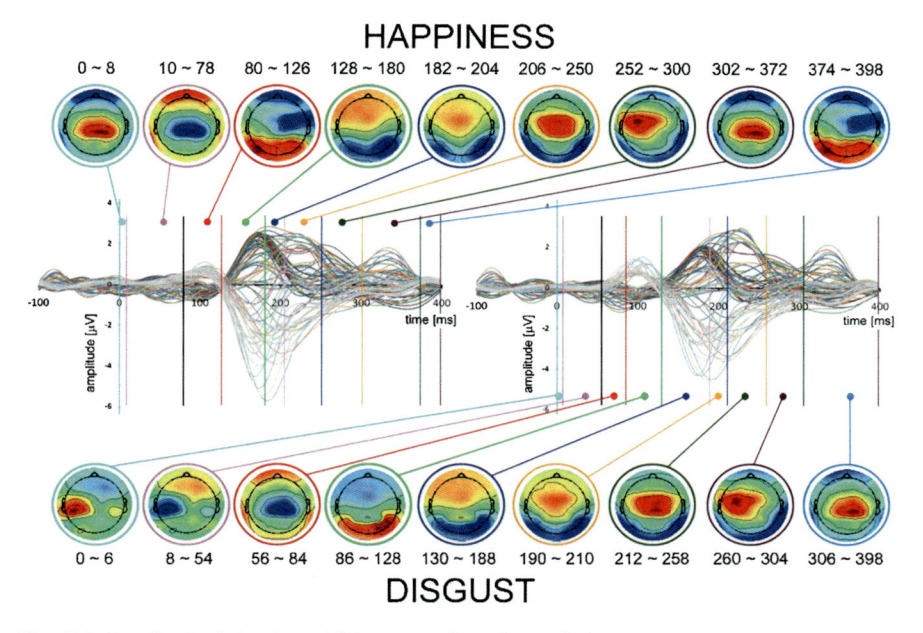

Fig. 7.6 Result of windowing with k-means clustering technique

Brain-Mind Research, University Hospital for Psychiatry, Zurich (http://www.uzh. ch/keyinst/loreta.htm). Detail for using the software is also available from the same link above.

7.5 Temporal Characteristics in the Recognition of Emotional Contain of Facial Expressions

Many studies have been performed to answer the question of how the brain processes the emotional contents in facial expression. But, still there are disagreements in temporal characteristics of the processing of facial emotions. To settle with the disagreements, we examined the brain signals that were evoked by visual stimuli of facial expressions using EEG. As known, facial expressions contain information about the type of the emotion as well as its intensity level. Therefore, to answer how the type and intensity level of emotions affect the brain activity, we parametrically controlled the intensity level, as well as the type of emotional content in facial expression. To elucidate the neural mechanisms related to the processing of these two parameters, we adopted morphed images in between neutral and the emotional ones (0, 10, 20, 40, 60, and 100 %). These percentages correspond to the scale level of morphed into the emotional images, where 0 % means neutral facial expression, and 100 % means full scale of emotion of facial

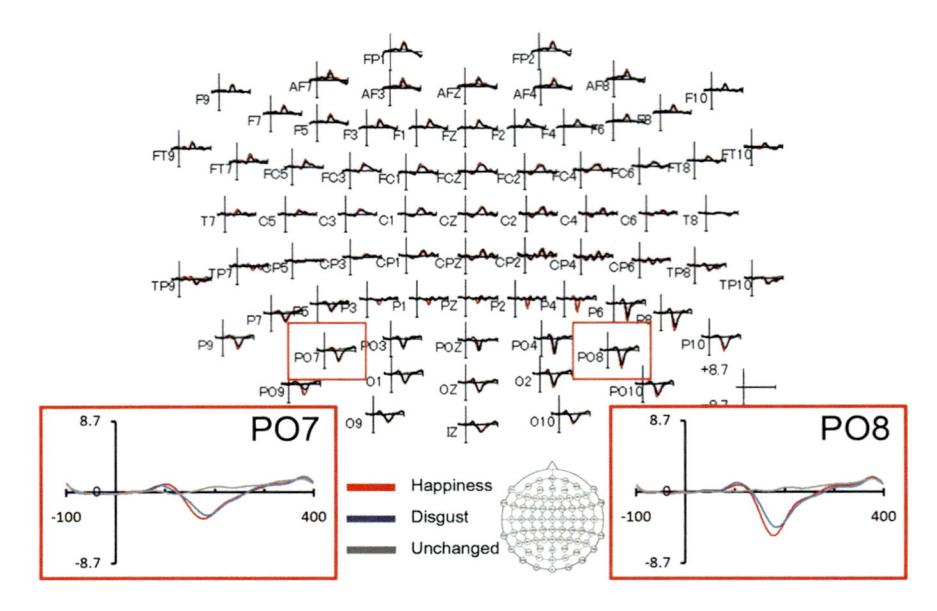

Fig. 7.7 Grand-averaged ERP waveform for the emotions happy (*red*) and disgusted (*blue*) and the unchanged face (*gray*) from 73 electrodes (*n* = 30). The arrangement of electrodes is based on a *top view*, and thus, the *top* corresponds to the anterior and the *bottom* to the posterior. Inserts are enlarged graphs at PO7 and PO8 locations (*left* and *right*, respectively)

expressions, whereas 10, 20, 40, and 60 % are the artificial images with morphed scale level of 10, 20, 40, and 60 % toward the full emotional of facial expressions (see Fig. 7.1). Subjective ratings in classifying the type and the intensity level of emotional contents in the presented facial expressions were the parameters in categorizing the temporal activities of the brain.

The best of EEG analyses is the precise temporal detection of brain activities regard to the presented stimuli in this ERP study. Through the data analyses, clear responses were observed in the posterior (occipital) and anterior (frontal) regions (see Fig. 7.7). In the posterior electrode locations, there appeared to be a positive deflection at around 100 ms followed by a negative deflection at around 170 ms (inset of Fig. 7.7). The anterior location response patterns counterbalanced those in the posterior locations. In positive valence, we focus in analyzing ERP data correspond to happy facial expression, meanwhile in the negative valence, we focus in ERP data of disgust facial expression. Using the k-means clustering analysis, from data of 30 subjects, we found four time-range windows of interest for each type of facial emotion (see Fig. 7.6). The time-range (post-stimulus onset) windows for happiness were (90–110 ms), (138–180 ms), (182–204 ms), and (206–230 ms), and those for disgust were (86–120 ms), (142–188 ms), (190–210 ms), and (212–258 ms). It is normal to have different result of time-range between the happiness and disgust because happiness and disgust are differently processed by the brain. We designated these four time-range windows as Window-1, Window-2,

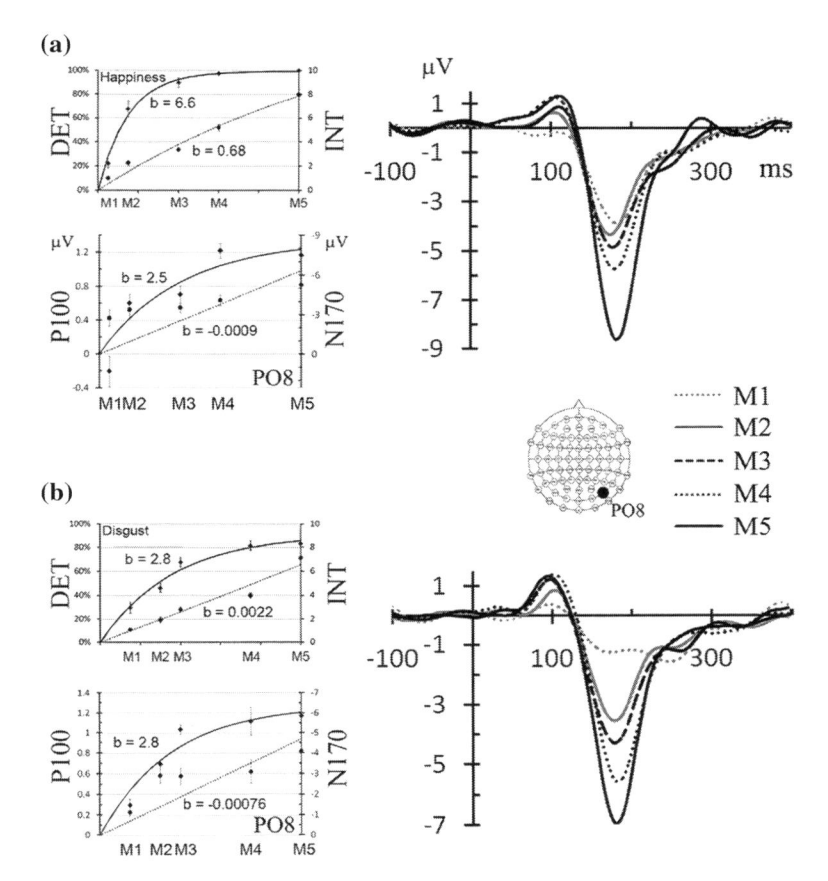

Fig. 7.8 Subjective rating data and the effects of emotion intensity on the ERP response. Averaged DET and INT scores during the experiment (*top left*). Grand-averaged waveform calculated by subtracting the ERP response to unchanged faces from that to happy faces with five different intensity levels; 10, 20, 40, 60, and 100 %, and designated as M1–M5 (*right*), respectively. At the posterior location PO8 and change in the mean voltage value depending on the intensity of the facial emotion (*bottom left*). The location of PO8 is shown in the central insert. **a** Data for happy facial expressions. **b** Data for disgusted facial expressions. *DET* correct detection of facial emotion; *INT* assessment of its intensity

Window-3, and Window-4, respectively. To identify the effect of the five emotion intensity levels on the ERP response, the grand-averaged waveforms were determined by subtracting the ERP response to the unchanged face from those to the happy face and disgusted face. Due to changes in intensity, the ERP signal changed its magnitude around 100 and 170 ms post-stimulus (see Fig. 7.8).

7.5.1 Window-1 (P100)

The ERP component within the first time-range Window-1 included the 100-ms post-stimulus onset and was designated as P100. To represent these components, the mean voltage values during Window-1 and Window-2 were calculated. The mean values were collected from all electrode locations, and the change in the value was compared with the change in the type of emotional content (DET) and the intensity level of emotional content (INT). By comparing the peak value among the electrode locations with the ratings (DET and INT), we found that the peak value of P100 is mostly similar to the subjective rating of DET compared with that of INT, especially in the frontal and posterior areas. For example, the peak value at the PO8 electrode location increased along with the increment of emotional level in the presented stimuli. However, when comparing this increment with that of subjective ratings (DET and INT), we found that the peak value at PO8 was significantly and strongly correlated with DET (DET, $r = 0.97$, $p = 0.01$) but not with INT (INT, $r = 0.86$, $p = 0.10$). Similar results were obtained for the emotion disgusted (solid line, lower left Fig. 7.8b; DET, $r = 0.98$, $p = 0.01$; INT, $r = 0.84$, $p = 0.10$).

To examine the between-subject variability, we calculated the number of subjects exhibiting a significant correlation between the ERP components and subjects' performance ($p < 0.05$) at each electrode location and then made a frequency map of the significant correlation. The right occipito-parietal locations showed a significant and consistent positive correlation between P100 and DET for both the emotions happy and disgusted (see Fig. 7.9, P100). Several other locations showed a significant correlation between P100 and DET or INT but with less consistency among subjects (see Fig. 7.9). We further compared the correlation coefficient between P100 and DET with that between P100 and INT for all 73 electrode locations. For the emotion happy, the DET value was higher than that of INT in seven of nine subjects (Wilcoxon signed rank test, $p < 0.05$). For the emotion disgusted, in three of nine subjects, the DET value was higher than that of INT ($p < 0.05$). These data suggest that P100 was more strongly correlated with DET than INT.

7.5.2 Window-2 (N170)

The ERP component within the second time-range Window-2 included the 170-ms post-stimulus onset and was designated as N170. The change in the mean voltage value was compared with the change in DET and INT. The N170 value at the PO8 location increased linearly with emotional intensity (dotted line, Fig. 7.8a). For the emotion happy, the PO8 value was significantly correlated with INT ($r = -0.99$, $p = 0.01$) but not DET ($r = -0.80$, $p = 0.10$; INT). Similar results were obtained

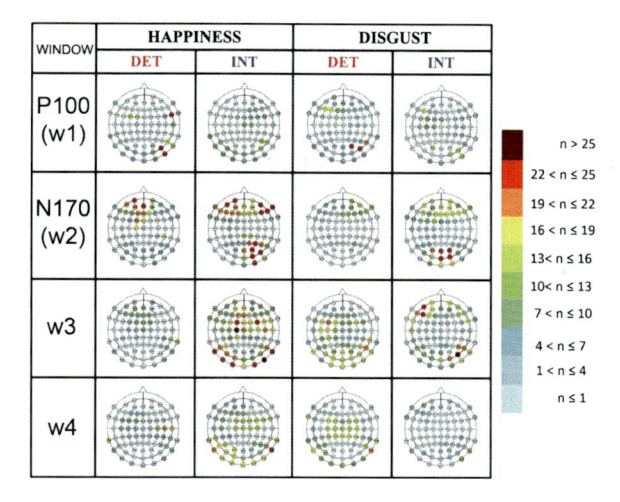

Fig. 7.9 Topographical maps of the electrodes showing the statistically significant correlation among subjects for each time-range window. The consistency in the significant correlation ($p < 0.05$) among subjects is represented by the number of subjects (n), indicated by color in the color bar. W1–W4 are the first to fourth time-range windows of interest for the emotion happy, (W1, 90–110 ms), (W2, 138–180 ms), (W3, 182–204 ms), and (W4, 206–230 ms) and for the emotion disgusted (W1, 86–120 ms), (W2, 142–188 ms), (W3, 190–210 ms), and (W4, 212–258 ms). *DET* correct detection of facial emotion; *INT* assessment of intensity

for the emotion disgusted (dotted line Fig. 7.8b; DET, $r = -0.86$, $p = 0.10$; INT, $r = -0.91$, $p = 0.05$).

As shown in Fig. 7.9 (N170), the right occipito-parietal locations showed a significant and consistent correlation between N170 and INT for both the emotions happy and disgusted. In addition, for the emotion happy, bilateral frontal locations showed a significant correlation between N170 and INT, and the left frontal locations showed a significant correlation between N170 and DET. For the emotion disgusted, the frontal locations also showed a significant correlation between N170 and INT but with less consistency among subjects. When we compared the correlation coefficient between N170 and DET with that between N170 and INT, the INT value was higher than that of DET ($p < 0.05$) in seven and eight of nine subjects for the emotions happy and disgusted, respectively. These data suggest that N170 was correlated more with INT than DET.

Both DET and INT affected ERP components. The magnitude of P100 sharply increased as the intensity of the facial emotion increased, and the P100 magnitude reached a plateau at less than half of the strongest intensity level. We demonstrated that the P100 magnitude was significantly correlated with DET. On the other hand, we failed to find any significant differences in the magnitude of P100 between the emotions happy and disgusted. These data suggest that the P100 is closely associated with the correct detection of facial emotion. Previous studies have reported that the facial emotion evokes brain activities at a very early stage of processing

(Pizzagalli et al. 1999b; Eger et al. 2003). In agreement with these studies, our data suggest that the brain activity evoked by the happy and disgusted faces occurred very early (100 ms) in the processing stage. Our data suggest that the P100 magnitude represents detection accuracy, but not the ability to distinguish these facial emotions. The detection of a facial emotion is probably a more primitive process than identification and needs less perceptual demand. Different neural mechanisms probably underlie these two processes.

However, in the present study, the subjects were required to answer the type of emotion, such as happy, angry, disgusted, etc., not just *'something emotional.'* In this sense, the subjects' subjective rating performance, DET, was not *'detection of facial emotion'* but *'identification of type of facial emotion.'* Because we adopted the stimuli exhibiting a similar DET–INT discrepancy, i.e., the emotions happy and disgusted, the similar profile in psychological data may cause our failure to find any significant difference in P100 magnitude between the emotions happy and disgusted. We might detect some significant correlation between P100 and the identification of type of emotion if we adopt stimuli exhibiting clearly different psychological profiles, such as the emotions happy and fearful.

7.5.3 Window-3 and Window-4

Similarly, the mean voltage value during Window-3 and Window-4 was calculated, and the change in the voltage value was compared with the change in DET and INT (see Fig. 7.9, W3 and W4). Similar to N170, the Window-3 value for the emotion happy at the occipital and parieto-frontal locations was significantly and consistently correlated with INT but not with DET. For the emotion disgusted, the value at the right posterior and left frontal locations was correlated with INT but not DET. The Window-4 value at the occipito-temporal location was correlated with INT for the emotion happy. The value at the occipital and parieto-frontal locations showed a significant correlation for DET or INT but with less consistency. Because of the less consistency among subjects, we did not extensively analyze the data in Window-4. We compared the correlation coefficient between the Window-3 value and DET with that between the Window-3 value and INT. For the emotion happy, the correlation with INT value was higher than that of DET in 27 subjects' data. For the emotion disgusted, data from five subjects indicated similar correlation with DET and with INT; fifteen data indicated higher correlation to DET compared with that of INT; meanwhile, the remaining ten data indicated that the correlation with INT is higher correlated than that with DET. Thus, by this finding, we can simply conclude that the Window-3 value was correlated with INT for the emotion happy but not for the emotion disgusted.

Based on these findings, we determined that DET occurred before INT. The system that detects facial emotions could be connected to the *'early-warning'* system that helps animals to survive by detecting potentially dangerous or threatening signals, such as fear and disgust (Morris et al. 1996; Phillips et al. 1997). Phased or

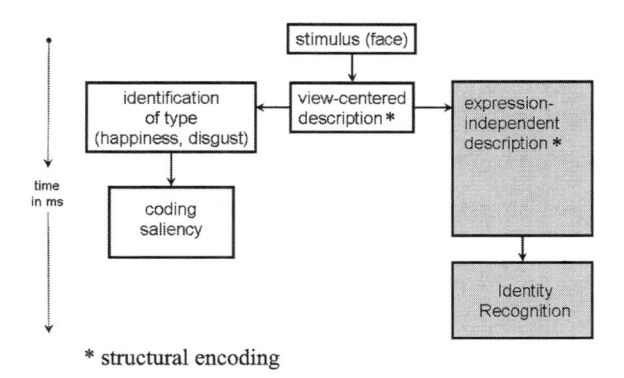

* structural encoding

Fig. 7.10 Two-phase model of emotion recognition. Phase 1 represents the initial monitoring process that codes saliency of incoming (structural encoded) facial information. In phase 2, the specific emotional content of faces (the type) is analyzed in emotion-specific recognition systems followed by the coding saliency of facial emotions. The *gray box* represents the very much simplified identity recognition system. *White boxes* represent the early step of facial emotion recognition systems or the expression-dependent description systems

serial-processing mechanisms in the brain may occur in relation to the processing of facial emotion (Adolphs 2002). Initially, the detection of facial emotion begins as early as 100 ms, and the detail is sufficiently constructed by around 170 ms post-stimulus to create distinguishing information. This phased mechanism enriches the functional-level model proposed by Bruce and Young (1986) as modified by Haxby et al. (2000) (see Fig. 7.10). Others have found processing components with a latency of more than 200 ms (Bobes et al. 2000; Carretié and Iglesias 1995). These components might be related to conceptual knowledge of the emotion that is signaled by the face, such as an interaction with gaze (Klucharev and Sams 2004), and face familiarity (Schweinberger et al. 2002). These later components have been localized by intra-cerebral electrode recording (McCarthy et al. 1999). Brain imaging studies have identified that the inferior occipital gyrus, the posterior fusiform gyrus, and the temporal poles are involved in facial processing (Haxby et al. 2000; Kanwisher et al. 1997; Nakamura et al. 2000). These findings suggest that a variety of brain regions function cooperatively to process facial emotion and that the activity in these regions is modulated by top–down and bottom–up signals. Further studies are necessary to elucidate the functions of the later components in the processing of facial emotion.

References

Adolphs R (2002) Neural systems for recognizing emotion. Curr Opin Neurobiol 12:169–177
Adolphs R, Tranel D (2003) Amygdala damage impairs emotion recognition from scenes only when they contain facial expressions. Neuropsychologia 41:1281–1289

Balconi M, Pozzoli U (2003) Face-selective processing and the effect of pleasant and unpleasant emotional expression on ERP correlates. Int J Psychophysiol 49:67–74

Benowitz LI, Bear DM et al (1983) Hemispheric specialization in nonverbal communication. Cortex 19:5–11

Bobes MA, Martin M, Olivares E, Valdes-Sosa M (2000) Different scalp topography of brain potentials related to expression and identity matching of faces. Brain Res Cogn Brain Res 9:249–260

Borod JC, Andelman F et al (1985) Channels of emotional expression in patients with unilateral brain damage. Arch Neurol 42:345–348

Bowers D, Bauer RM et al (1985) Processing of faces by patients with unilateral hemisphere lesions: I. Dissociation between judgments of facial affect and facial identity. Brain Cogn 4:258–272

Bruce V, Young A (1986) Understanding face recognition. Br J Psychol 77:305–327

Bruce V, Young AW (1998) A theoretical perspective for understanding face recognition. In: Young AW (ed) Face and mind. Oxford University Press, Oxford, pp 96–131

Calder AJ, Keane J et al (2000) Facial expression recognition by people with mobius syndrome. Cogn Neuropsychol 17:73–87

Carretié L, Iglesias J (1995) An ERP study on the specificity of facial expression processing. Int J Psychophysiol 19:183–192

Darwin C (1872) The expression of the emotions in man and animals. Digireads.com Publishing, Stilwell

Delorme A, Makeig S (2004) EEGLAB: an open source toolbox for analysis of single-trial EEG dynamics including independent component analysis. J Neurosci Methods 134:9–21

Eger E, Jedynak A, Iwaki T, Skrandies W (2003) Rapid extraction of emotional expression: evidence from evoked potential fields during brief presentation of face stimuli. Neuropsychologia 41:808–817

Ekman P (2003) Emotions revealed: recognizing faces and feelings to improve communication and emotional life. Times Books, New York

Ekman P, Friesen WV (1976) Measuring facial movement. Environ Psychol Nonverbal Behav 1(1):56–75

Ellis HD, Young AW (1998) Faces in their social and biological context. In: Young AW (ed) Face and mind. Oxford University Press, Oxford, pp 67–96

Field TM, Woodson R et al (1982) Discrimination and imitation of facial expressions by neuronates. Science 218:179–181

Finger S (1994) Origins of neuroscience: a history of explorations into brain function. Oxford University Press, Oxford, pp 41–42

Fredikson M, Wik G et al (1995) Functional neuroanatomy of visually elicited simple phobic fear: additional data and theoretical analysis. Psychophysiology 32(1):43–48

Fuchs M, Kastner Jn, Wagner M, Hawes S, Ebersole JS (2002) A standardized boundary element method volume conductor model. Clin Neurophysiol 113:702–712

Haxby JV, Hoffman EA, Gobbini MI (2000) The distributed human neural system for face perception. Trends Cogn Sci 4:223–233

Heberlein AS, Adolphs R et al (2003) Effects of damage to right-hemisphere brain structures on spontaneous emotional and social judgments. Polit Psychol 24(4):705–726

Heller W, Nitschke JB et al (1998) Lateralization in emotion and emotional disorders. Curr Dir Psychol Sci 7(1):26–32

Jung TP, Makeig S, Humphries C, Lee TW, McKeown MJ, Iragui V, Sejnowski TJ (2000a) Removing electroencephalographic artifacts by blind source separation. Psychophysiology 37:163–178

Jung T-P, Makeig S, Westerfield M, Townsend J, Courchesne E, Sejnowski TJ (2000b) Removal of eye activity artifacts from visual event-related potentials in normal and clinical subjects. Clin Neurophysiol 111:1745–1758

Junghofer M, Bradley MM et al (2001) Fleeting images: a new look at early emotion discrimination. Psychophysiology 38:175–178

Kanwisher N, McDermott J, Chun MM (1997) The fusiform face area: a module in human extrastriate cortex specialized for face perception. J Neurosci 17:4302–4311

Klucharev V, Sams M (2004) Interaction of gaze direction and facial expressions processing: ERP study. NeuroReport 15:621–625

Lancaster JL, Woldorff MG, Parsons LM, Liotti M, Freitas CS, Rainey L, Kochunov PV, Nickerson D, Mikiten SA, Fox PT (2000) Automated Talairach atlas labels for functional brain mapping. Hum Brain Mapp 10:120–131

Lane RD, Chua PML et al (1998) Common effects of emotional valence, arousal and attention on neural activation during visual processing of pictures. Neuropsychologia 37:989–997

Lehmann D, Skrandies W (1984) Spatial analysis of evoked potentials in man—a review. Prog Neurobiol 23:227–250

Mazziotta J, Toga A, Evans A, Fox P, Lancaster J, Zilles K, Woods R, Paus T, Simpson G, Pike B, Holmes C, Collins L, Thompson P, MacDonald D, Iacoboni M, Schormann T, Amunts K, Palomero-Gallagher N, Geyer S, Parsons L, Narr K, Kabani N, Le Goualher G, Boomsma D, Cannon T, Kawashima R, Mazoyer B (2001) A probabilistic atlas and reference system for the human brain: international consortium for brain mapping (ICBM). Philos Trans R Soc Lond B Biol Sci 356:1293–1322

McCarthy G, Puce A, Belger A, Allison T (1999) Electrophysiological studies of human face perception. II: response properties of face-specific potentials generated in occipitotemporal cortex. Cereb Cortex 9:431–444

Michel CM, Henggeler B, Lehmann D (1992) 42-channel potential map series to visual contrast and stereo stimuli: perceptual and cognitive event-related segments. Int J Psychophysiol 12:133–145

Morris JS, Frith CD, Perrett DI, Rowland D, Young AW, Calder AJ, Dolan RJ (1996) A differential neural response in the human amygdala to fearful and happy facial expressions. Nature 383:812–815

Nakamura K, Kawashima R, Sato N, Nakamura A, Sugiura M, Kato T, Hatano K, Ito K, Fukuda H, Schormann T, Zilles K (2000) Functional delineation of the human occipito-temporal areas related to face and scene processing. A PET study. Brain 123:1903–1912

Natale M, Gur RE et al (1983) Hemispheric asymmetries in processing emotional expressions. Neuropsychologia 21:555–565

O'Leary JL (1970) Hans Berger on the electroencephalogram of man. The fourteen original reports on the human electroencephalogram. Science 168:562–563

Pascual-Marqui RD (2002) Standardized low-resolution brain electromagnetic tomography (sLORETA): technical details. Methods Find Exp Clin Pharmacol 24(Suppl D):5–12

Pascual-Marqui RD, Michel CM, Lehmann D (1995) Segmentation of brain electrical activity into microstates: model estimation and validation. IEEE Trans Biomed Eng 42:658–665

Philipps ML, Bullmore ET et al (1998) Investigation of facial recognition memory and happy and sad facial expression perception: an fMRI study. Psychiatry Res: Neuroimaging 83:127–138

Phillips ML, Young AW, Senior C, Brammer M, Andrew C, Calder AJ, Bullmore ET, Perrett DI, Rowland D, Williams SC, Gray JA, David AS (1997) A specific neural substrate for perceiving facial expressions of disgust. Nature 389:495–498

Pizzagalli D, Koenig T et al (1999a) Affective attitudes to face images associated with intracerebral EEG source location before face viewing. Cogn Brain Res 7:371–377

Pizzagalli D, Regard M, Lehmann D (1999b) Rapid emotional face processing in the human right and left brain hemispheres: an ERP study. NeuroReport 10:2691–2698

Reuter-Lorenz P, Davidson RJ (1981) Differential contributions of the two cerebral hemispheres to the perception of happy and sad faces. Neuropsychologia 19:609–613

Robinson RG, Downhill JE (1995) Lateralization of psychopatology in response to focal brain injury. In: Davidson RJ, Hugdahl K (eds) Brain asymmetry. MIT Press, Cambridge, pp 693–711

Sato W, Kochiyama T et al (2001) Emotional Expression boosts early visual processing of the face: ERP recording and its decomposition by independent component analysis. NeuroReport 12:709–714

Schwartz GE, Davidson RJ et al (1975) Right hemisphere lateralization for emotion in the human brain: interactions with cognition. Science 190(4211):286–288

Schweinberger SR, Pickering EC, Jentzsch I, Burton AM, Kaufmann JM (2002) Event-related brain potential evidence for a response of inferior temporal cortex to familiar face repetitions. Brain Res Cogn Brain Res 14:398–409

Talairach J, Tournoux P (1988) Co-planar stereotaxic atlas of the human brain: 3-dimensional proportional system—an approach to cerebral imaging. Thieme Medical Publishers, New York

Utama NP, Takemoto A, Koike Y, Nakamura K (2009) Phased processing of facial emotion: an ERP study. Neurosci Res 64:30–40

Printed by Publishers' Graphics LLC
MLSI140312.15.20.120